The Doctor's 30-Day Cholesterol BLITZ

Leslie C. Norins, M.D., Ph.D.
The Physician Who Healed Himself
and
Rainey Norins
His "Kitchen Coach"

ADVANCED HEALTH INSTITUTE, INC.
851 Fifth Avenue North, Suite 301
Naples Florida 33940
1-800-508-2582

THE DOCTOR'S 30-DAY CHOLESTEROL BLITZ

Published by

Advanced Health Institute, Inc.
851 Fifth Avenue North, Suite 301
Naples, FL 33940
1-800-508-2582

ISBN 1-888266-00-7

Printed in the United States of America
1st printing 1996

Book design by Stephen J. Gray

Important Notice

This monograph conveys much useful information about cholesterol control, but of necessity it is in generalized form. I cannot know the specific situation and clinical details concerning every reader and in no way is this book intended to replace, countermand or conflict with the advice given to you by your own physician or other health professional. Please consult your own doctor before undertaking any advice or treatment presented herein, and do not cease any current medications or treatments without your doctor's approval. Neither the author nor the publisher make any representation or warranty of any kind whatsoever regarding the effectiveness or appropriateness of this program, principles, or information for any individual.

Foreword

This report was written to help you successfully counterattack a high cholesterol problem--fast! Frankly, I got sick and tired of professionals' regarding cholesterol as a "chronic" affliction, slowly worsening, needing ever-increasing pills and diets that are more and more unpleasant.

Please notice this monograph is a lot thinner than a typical book. I wanted to get right down to reducing cholesterol and not waste a lot of paper and time on side excursions such as lipid chemistry, the way the heart works, anatomic drawings of blood vessels, etc.

Let me admit right up front that I challenge the notion that cholesterol must be a permanent problem for you. If you accept this propaganda, you fall into the trap of thinking there's nothing much that can be done. So you dutifully swallow pills every day and make a few half-hearted food adjustments -- and wander around under a constant death threat from fat clogging your heart's blood vessels.

But it's time to talk about counterattack. My own 30-day success inspired me -- and everybody who heard about it wanted the details, thus we wrote this report.

As you will read, the core concept of my victory program is the BLITZ. I use everything and anything I can to lower cholesterol -- simultaneously. And I refuse to be drawn into the artificial debate between "pills" and "natural remedies". I believe you should get all the helpful items you can onto your team, and then -- with all working together and concurrently -- BLITZ your cholesterol problem into submission.

Remember, do not try one ingredient for six months, then another for six, etc., in a slow water-torture progression. Instead, bring into the fight as many of my weapons as you feel comfortable with and employ them all at the same time.

It's going to be a lot easier than you might have guessed. First, you'll have the proven motivation of knowing it can be done quickly, and that it worked for me. Secondly, with the clever substitutions of my Near 'Nuff foods (which you'll read about), you won't even feel deprived. (How do "hamburgers", "loaded baked potatoes", "chocolate milkshakes", and "BLT" sandwiches sound?)

I'm 99% certain you'll quickly get your cholesterol into the normal range with this program. Just have an open mind and give me your enthusiasm and cooperation. I'll thank you and your heart will thank you.

Leslie C. Norins, M.D., Ph. D.

About The Authors

Dr. Leslie C. Norins has had a 23-year entrepreneurial career in health-care publishing. He founded and directed the world's largest medical newsletter company, and is currently president of Global Success Corporation, a medical publishing company. Dr. Norins received his A.B. from Johns Hopkins University, his M.D. from Duke University Medical School, and his Ph.D. from the University of Melbourne (Australia), where he was the personal student of the Nobel Laureate, Sir MacFarlane Burnet. Dr. Norins is a Fellow of the Infectious Disease Society of America, has served on committees of the National Institutes of Health and the World Health Organization, has been on the Board of Directors of the Newsletter Publisher's Association, and has been selected for *Who's Who in America*.

Rainey Norins, Dr. Norins' wife, is president of Advanced Health Institute, which publishes consumer health information. Before this publishing phase of her career, she directed several business enterprises in other fields. Her love of cooking and healthy nutrition started early; as a young person she took part in a rural 4-H program, and then studied home economics in college. The Norins' live and work in Naples, Florida.

Table of Contents

Introduction: How The BLITZ Was Born

Not long ago -- in mid-1995, to be exact -- I was forced to acknowledge that my health was under siege.

As I sat in my doctor's office, listening while he somberly recited the disturbing results of my recent medical tests, there was no way I could rationalize away my personal health crisis. My blood cholesterol reading had soared to a stratospheric 357. My triglycerides had surged, too, up to a startling 1,154. My blood pressure was also too high -- 144/90 -- even though I was taking the highest recommended dose of the medication prescribed to control it.

As a physician, I knew exactly what these numbers meant: Unless I took action to regain control of my health -- unless I really got serious about managing my cholesterol and triglycerides -- I was on the fast track toward a heart attack and a shortened lifespan.

That's when my wife Rainey and I began working together to create The Doctor's 30-Day Cholesterol BLITZ -- a revolutionary program whose "Nuggets" of lifestyle changes literally transformed my health in only a month. My test results improved so quickly and dramatically that even my doctor was stunned by my rapid progress.

The BLITZ is a program that anyone can adopt, no matter what his or her health status. As you'll discover in this monograph, it is a comprehensive, multifaceted plan that attacks high cholesterol from many directions. Not just by slashing the fat in the diet, not just by working out regularly, but by "blitzing" with the throttle wide open, utilizing every available approach at once. Bringing your cholesterol under control and improving your overall health is actually simpler, and can happen much faster, than you probably believe. In fact, if good health is important to you, I urge you to give this program a try, and see what a difference it can make.

In the following pages, I'll describe the BLITZ in detail. You'll get every piece of information you need to open fire upon your own high cholesterol level. You won't be adopting strategies one after the other in a slow, methodical manner, but you'll be staging a full-frontal assault on those habits that have been sabotaging your chances at good health.

But first, let me begin by telling you a little more about my own experience, and how the BLITZ was created. You may find yourself identifying with many parts of my story, and I hope you'll be inspired to make the same kinds of health-enhancing changes in your own life.

■ ■ ■ ■

Chapter 1

Time For A Cholesterol Change

For years, my lifestyle was not much different than that of millions of other allegedly health-conscious men and women. During that time I had been casually attempting to follow a low-fat diet and even managed to convince myself I was eating healthy foods, reading labels and walking briskly for exercise. In reality, I probably ate more fatty foods than I should have. I weighed too much and exercised too little to be effective. So I shouldn't have been surprised when my doctor gave me the sobering results of my medical tests, which were taken on the eve of the BLITZ:

Table 1

Pre-BLITZ Lipid Levels

	June 15	Upper Limit of "Normal"[2]
Total cholesterol	357[1]	200
Triglycerides	1154	400
LDL ("bad") cholesterol	NC[3]	130

Notes:

1. All numbers are "milligrams per deciliter".

2. Normal values are included for comparison.

3. NC = not calculable. LDL cholesterol is calculated by using a formula, but the calculation is not really valid when triglycerides are over 400, as they were on this occasion.

I had been taking medication for high blood pressure for 14 years, so I got the message some time ago that I needed to be more attentive to my health. But perhaps I was still clinging to the reality of my twenties and thirties, when I could eat fattening foods -- from pizzas to pastries -- with impunity, and never waver from a weight of 160 pounds. When my doctor pointed to the new "total cholesterol" figure of 357, however, and explained the urgency of starting on a cholesterol-lowering medication called Mevacor, it was time for some serious soul-searching. Like a rubber band that loses some of its resiliency with constant use, my body just wasn't as capable of rebounding from my overindulgences as it had been two decades earlier. I realized I had to make some changes, and I needed to stick with them.

Still, I was anxious about what might lay ahead. What did I really have to look forward to? Months of hard work, producing only a five percent decline in my cholesterol and triglyceride levels? A lifetime of taking Mevacor? Would I be "leaving life as I knew it," giving up all the tastes and flavors of the foods I loved? Would I be munching on twigs and berries, tree barks and leaves, for the rest of my life?

Let's face it: Most of us don't have the patience for slow, steady change, even if it's in a desirable direction. I wanted (and felt I needed) rapid and substantial improvements to stay motivated. So rather than inching my way to good health, I was determined to take a giant leap forward. However, even though I was frightened by my high cholesterol and triglyceride readings in those early days, I didn't know where to begin.

This may surprise you, since in addition to an M.D., I also have a Ph.D. in medical research, and I studied in the laboratory of a Nobel prize winner. But there's a mass of information out there with evangelists from all sides -- "pills", "natural" vitamins & minerals, food plans, exercise freaks, meditators, etc. So I can sympathize if you've been a little confused, too. How could I boil it all down and learn what to do to bring my test results down into a more comfortable zone?

That's when Rainey and I began our own cram course on cholesterol. We raided local bookstores, carrying home armfuls of books on cholesterol reduction and heart disease. We read them quickly, looking for "Nuggets" to make part of our program. One author favored a diet rich in oat bran. Another recommended vitamin supplements. Still another advised eating lots of beans and grains. We found many strategies that made sense to us, some based on solid scientific evidence, and others that seemed reasonable although they might not yet be proven nor widely known.

Having directed a research laboratory at the federal Centers for Disease Control in the early years of my career, where I specialized in infectious diseases, I remained driven by a philosophy that had served me well there: When confronted with an epidemic, you don't limit yourself to a single weapon for the battle -- you use everything at your disposal. So in fighting my personal health nemesis (high cholesterol), my goal was to employ every possible strategy simultaneously (as long as there was no evidence that they could do harm when grouped together), believing this approach could produce more rapid results.

What's more, it might take advantage of a synergistic effect, with each "weapon" enhancing the effect of others. So rather than employing just one Nugget, Rainey and I decided that using many of them together -- from the traditional to the more obscure -- was the approach we'd take. For example, when the value of apple pectin became clear, I made it part of our arsenal. Once the importance of physical activity became undeniable, I made a specific commitment to physical exercise and weight training at our home. When the evidence favoring garlic appeared attractive, I welcomed it to the team.

We called this many-items-at-a-time approach "The BLITZ". It is modeled on the military philosophy of mounting an offensive from all angles -- on the ground and in the air -- to overpower the enemy. Or the football strategy of surprising the quarterback with a rush by every available defensive linebacker at once. We figured a multi-faceted BLITZ would have the best chance of vanquishing my high cholesterol levels quickly.

In just days, Rainey and I were ready to start the BLITZ. And once we did, the results were astounding. The first change I noticed was in my blood pressure, which I could measure at home. I really wasn't expecting much improvement, particularly in those first few days. But after just a week, my blood pressure was noticeably declining. Then, when I returned to my doctor's office for blood tests a month after the BLITZ began, the full scope of the changes became clear. My total cholesterol readings had plunged 57 percent! And my triglycerides had gone into a real free-fall, declining 79 percent!

Meanwhile, I began to wean myself off of my blood pressure medication (a beta blocker called Tenormin), over a 30-day period, gradually reducing the dosage from 100 mg. a day down to zero. After 14 years on this drug, my blood pressure settled in at a healthy 132/88 level without medication. I also lost 14 pounds of weight. I went off the cholesterol-lowering drug as well.

> *Do not discontinue or lower your own medications without your doctor's approval. I took this action in my own care based on my experience and knowledge as a physician, and believed my own doctor would concur -- which he did.*

My doctor was so astonished that he wrote on my chart, "It has been a dramatic turnaround." The improvements were so impressive, in fact, that I decided to stick to a modified BLITZ, which I term "maintenance" BLITZ, after that first 30 days. Despite my initial anxieties about having to survive on twigs and berries, my taste buds never felt deprived. Rainey and I decided to stay on the program indefinitely -- and today, many months later, the BLITZ has become part of our lives.

Here are the improvements I experienced during the first 30 days of the BLITZ:

Table 2

Results of the BLITZ After 30 Days

	BLITZ Day 1	BLITZ Day 30
Blood Lipids:		
Total cholesterol	357	155
Triglycerides	1154	248
LDL ("bad") cholesterol	NC[1]	73
Other Factors:		
Weight	195	181
Blood pressure	144/100[2]	132/88[3]

Notes:

1. See note 3 in Table 1.

2. With medication.

3. Without medication.

A CHOLESTEROL PRIMER

As disturbing as my high cholesterol levels were, I knew I wasn't alone. In fact, 95 million American adults (52 percent of the adult population) have blood cholesterol values of 200 or higher. "Desirable" is less than 200.

So what is cholesterol? It is a waxy, fat-like substance (also called a "lipid" or blood fat) which, in small amounts, is needed by the body to form cell membranes and certain hormones. But if there's too much cholesterol in your bloodstream -- a factor influenced by the amount of fat and cholesterol you eat -- fatty deposits can accumulate on the walls of the coronary arteries (which supply blood to the heart). As these "plaques" clog and narrow the arteries, they can interfere with blood flow, setting the stage for a heart attack.

In general, as your cholesterol level climbs, so does your risk of cardiovascular disease. Likewise, studies show that for every 1 percent decline in your cholesterol level, you can cut your chances of a heart attack by 2 percent; thus, if you can reduce your cholesterol reading by 20 percent -- say, from 250 to 200 -- you can slash your risk of heart problems by 40 percent!

Cholesterol is transported through the bloodstream by hitching rides on carriers called lipoproteins -- specifically LDL (low-density lipoproteins) and HDL (high-density lipoproteins).

HDL, often called "good" cholesterol, is the Good Samaritan in your bloodstream, actually kidnapping cholesterol from the arteries and shipping it out of the body. On the other hand, LDL ("bad") cholesterol is the mischief-maker, dumping cholesterol into the artery walls, and accelerating the plaque buildup that can clog your blood flow.

Just what should your cholesterol level be? Here's how the National Cholesterol Education Program classifies cholesterol readings:

Less than 200 -- desirable

200-239 -- borderline high

240 and above -- high

Your LDL level should be 130 or less. HDL goes the other way; the more of it the better (within reason); the minimal reading should not be lower than 35.

One other important point: I also had very high triglyceride levels. Triglycerides are another type of lipid, and are believed to contribute to your risk of heart disease. A triglyceride level below 200 is considered desirable, while readings above 400 are high. Fortunately, many of the Nuggets you'll find in this monograph not only slash away at your elevated cholesterol, but also at your high triglycerides.

A quick tip to help you remember which cholesterol is which:
HDL is the "Healthy" cholesterol, while
LDL is the "Lousy" cholesterol.

■ ■ ■ ■

Chapter 2

The Nutritional Nuggets: Basics Of The BLITZ

As we begin to describe the "Nuggets" in The Doctor's 30-Day Cholesterol BLITZ, let's start with the nutritional recommendations. Again, the intent of our program is to bombard the cholesterol problem intensively, attacking it from many directions simultaneously. Rainey and I searched the literature and scanned supermarket and health-food-store shelves, looking for ways to transform our diet into an ally in the war on cholesterol. Here are the shifts in food choices that we've discovered -- and that have become the nutritional Nuggets of the BLITZ:

Switch to low-fat and non-fat eating whenever possible

By cleaning up your act with lower-fat foods, you'll also help clean up your coronary arteries. At first glance, the shift to reduced-fat eating may sound difficult and painful; I certainly thought it would be. But it really isn't.

When it comes to cholesterol, your #1 target needs to be saturated fat, which raises cholesterol levels more than any other component of your diet. Be particularly wary of the major sources of saturated fat such as butter, cheese, whole milk, ice cream, and red meat. A few vegetable fats -- found in tropical oils like coconut, palm and palm kernel oil -- are also brimming with saturated fat, and thus they, too, need to be eaten with care.

What should you be consuming instead? Replace part of the saturated fat with unsaturated fat such as polyunsaturates (in safflower, corn, soybean, sesame and sunflower oils) and monounsaturates (in olive and canola oil). In moderation, these fats can actually reduce your total and LDL cholesterol readings. However, you shouldn't overconsume any types of fat, even those that are unsaturated. Remember, a little goes a long way with these oils since their calorie content is off the chart.

As you continue to read through these Nuggets, you'll find other suggestions on foods that can take the place of saturated fats. In fact, there are plenty of healthy (and tasty) reduced-fat choices available, and you'll never find yourself feeling deprived. For example, to help knock the fat out of your diet, select skim milk instead of low-fat or whole milk, and non-fat yogurt rather than the whole-milk variety. If you're a meat eater, rely on leaner choices like skinless chicken, and cut way down on serving sizes (3-4 oz. per meal). When buying cheese, look for those lower in fat or no-fat. And pare down these portions too; by cutting cheese down to size, you'll keep your coronary arteries from paying the price. Feast on fruit and vegetables, and retrain your sweet tooth to crave low-fat frozen desserts or fresh fruit. Rainey and I cook with non-stick pans (with flavored Pam spray), and avoid frying altogether. In this monograph, you'll find many suggestions like these for forgoing fat without sacrificing flavor and satisfaction.

By the way, most health agencies advise limiting your fat intake to less than 30 percent of the calories you consume. But I'm not the type of person who has the patience to weigh and measure every bite of food, and to dine with a calculator at my side. Frankly, I find it easier and just as effective to simply get rid of as much saturated fat in my diet as I can. By being conscientious about finding and discarding it, I'm keeping my fat intake right where I want it.

Eat a largely vegetarian diet

Before Rainey and I created The Doctor's 30-Day Cholesterol BLITZ and made it part of our lives, we had never been big meat eaters. Yes, we occasionally had red meat at dinner, and we ate chicken as well (Eating six pieces of Rainey's chicken, one after another, I used to tell myself, "It's OK; at least I'm not eating steak!"). Even so, we probably ate less meat than many Americans do.

When we began the BLITZ, however, we switched to an almost entirely vegetarian diet as a way to get rid of as much saturated animal fat as possible. At the same time, we also eliminated saturated vegetable fats like coconut and palm oils.

If your own dinner plate doesn't seem complete without a slab of steak or some finger-lickin' fried chicken, then it's time for a new mind-set. Early on, I couldn't picture vegetables as anything other than a side dish, and I truly believed that my vegetarian friends were missing out on something. In a sense, they were: They were missing out on high cholesterol levels and a greater likelihood of a heart attack! When Dr. Dean Ornish of the University of California, San Francisco, placed 28 people on a low-fat, vegetarian diet, 82 percent of them had less clogging of their coronary arteries after a year. In a separate study of vegetarian Seventh-Day Adventists, their LDL cholesterol was an average of 38 percent lower than a group of meat-eaters. At the same time, vegetarians are less likely to develop high blood pressure, kidney stones, gallstones and obesity.

So even if you can't fathom living on vegetables right now, keep an open mind. Although at first I couldn't imagine surviving without meat loaf and hamburgers, I found many other ways to entertain my taste buds. I urge you to try an almost meatless diet for 30 days, and you'll be surprised to find (as I did) that very-low-fat, vegetarian-based dishes can be a very satisfying alternative to lamb chops and ribs. Fill your plate with complex carbohydrates -- foods like vegetables, fruits, breads, pasta, rice, beans and potatoes. But be careful: If you prepare and serve these foods with butter, cream and rich sauces -- which are high in saturated fat -- you'll sabotage all your good intentions. Give your veggies some heat, steaming them to keep them nutrient-rich and crunchy. Refer to page 27 for condiment ideas on giving them flavor.

Also, we discovered that certain brands of frozen "veggie burgers", made from flavored and textured soy protein, or grains, taste almost identical to their real meat counterparts -- especially when put in a hamburger bun with lettuce, tomato and onion, and spread with catsup (ketchup) and mustard.

By the way, after the first 30 days of the BLITZ, an occasional three-ounce portion of fish or chicken (about the size of a deck of cards) is acceptable in the "maintenance" phase, but don't overdo it: Limit your intake to no more than three times a week. Choose light chicken meat rather than dark to minimize your fat intake, and also be sure to remove the skin from the chicken; by peeling off the skin, you'll remove half of the saturated fat in a typical piece of chicken. And when choosing seafood, focus your appetite on fish with lots of polyunsaturated omega-3 fatty acids (see box on page 16).

SHOULD YOU GET HOOKED ON FISH?

Have you ever wondered why the Greenland Eskimos, who eat a high-fat diet (40 percent of calories from fat), have a very low rate of heart disease? To find the explanation, look no further than the buckets of cold-water fish they catch and consume.

While red meat has deservedly earned a real nutritional black eye, seafood has a growing fan club, with many doctors and nutritionists leading the cheering. For starters, fish is low in saturated fat. But just as important, it contains omega-3 fatty acids, which are highly unsaturated fats that can actually take a bite out of your cholesterol count and triglyceride reading, while interfering with blood clotting in the vessels (thus cutting the risk of heart attacks). No wonder many experts now recommend eating fish two to three times a week to help fill the red-meat void.

Not all seafood is brimming in omega-3's, however. To avoid flirting with heart disease, choose the best sources: salmon, tuna, herring, mackerel, sardines, and trout. And to keep fish a healthy dish, rely on cooking methods like poaching, steaming, and grilling.

Fill up with fiber

By definition, fiber is any part of a plant that our bodies can't digest and absorb. You'll find fiber in many of the foods I've already mentioned in the vegetarian-diet section above. Thus, by making the switch away from meat, you'll begin to add more fiber to your diet.

There are two types of fiber -- soluble and insoluble. While the latter is important in a well-rounded diet (it can reduce your risk of colon and other types of cancer), soluble fiber is what we're most interested in when cholesterol reduction is the goal. Soluble fiber acts like a sponge, soaking up fats and cholesterol and transporting them out of the body before they can

create havoc for your heart. A study at the Stanford University School of Medicine found that eating a diet rich in soluble fiber (15 grams a day) decreased total cholesterol by 8.3 percent and LDL cholesterol by 12.4 percent in only one month.

Excellent sources of soluble and insoluble fiber include citrus fruits, dried beans, peas, rice, carrots, pasta, oat bran and barley. In fact, just about all types of fruits, vegetables and grains -- from apples to zucchini -- will give your diet a fiber boost. Here are just three fiber-rich options to consider:

Beans can be used in so many ways -- in everything from stews to salads to soups -- and it doesn't take many of them to help corral your cholesterol. Researchers at the University of Kentucky fed 1.5 cups a day of navy and pinto beans to men with high cholesterol; in just three weeks, their total cholesterol had plummeted an average of 56 points! Use beans as a substitute for beef in some of your favorite recipes, such as Mexican dishes.

Brown rice is not only full of fiber, but it contains rice-bran oil (which is rich in cholesterol-controlling polyunsaturated and monounsaturated fats). Researchers at the University of Massachusetts have also identified compounds in rice called "unsaponifiables" that interfere with cholesterol absorption. I recommend brown rice over white, since in the latter, the bran has been discarded and the oil removed.

Barley isn't widely used in the U.S., overshadowed by other grains like wheat and rice. But not only is barley a good source of soluble fiber, it contains polyunsaturated oils (called tocotrienols) that can help batter your cholesterol into submission. In a study at Montana State University, men and women with high cholesterol levels consumed barley flour added to bran muffins, breads and breakfast cereals; after six weeks, their total and LDL cholesterol levels fell 12 and 24 points, respectively. Look for barley in the pasta and rice aisles of your supermarket, and substitute it in recipes that call for rice. (It is also great to add to vegetarian chili for extra thickening, plus it is wonderful in most vegetarian soups.)

■ ■ ■ ■

Chapter 3

Near 'Nuff* Foods: Cholesterol-Cutting Without Deprivation

* "Near 'Nuff" is a Trademark of Advanced Health Institute, Inc.

If you're like me, you might be shuddering at the thought of adopting a dietary program that calls for the elimination of many of your favorite foods. I love many high-fat dishes (such as hot fudge sundaes) as much as anyone else. And as I've already mentioned, I was unwilling to swear off my favorite tastes and flavors in exchange for lower cholesterol levels (even though subconsciously I knew that I might have to in order to achieve my overall health goals). But much to my surprise, once Rainey and I began shopping carefully and conducting our personal taste tests of low-fat products and recipes, we found no need to give up much at all. Food science has found ways to mimic the tastes of many familiar foods -- but without very much of the harmful fat, sugar or calories to which we're accustomed. With The Doctor's 30-Day Cholesterol BLITZ, you can still eat foods that look and even taste fattening -- while consuming very little or no fat.

To avoid eating tree bark for breakfast, and leaves and twigs for lunch and dinner, Rainey and I searched for "crutches" -- what I call "near enough" foods: To our surprise, we found dozens of these alternatives that nearly replicated the tastes that I was in love with (I liked this concept so much that I even trademarked the term "Near 'Nuff"). These health-promoting substitutes helped me tremendously in making the transition to cholesterol-lowering eating.

Here are just a few of the Near 'Nuff substitutions that Rainey and I have discovered (others appear in the box on page 24):

Look for foods made with soy protein that are designed to imitate meat. Many soy-based frankfurters, veggie burgers and "chicken" patties look and taste like the real thing. They tend to be low in fat, but even the fat that they do contain is of vegetable origin and there's never cholesterol in vegetables. Rainey can't believe how often I want our newly discovered Near 'Nuff versions of "hot dogs" and "hamburgers" -- treating them as an entree and surrounding them with vegetables; of course, *everything* on the plate is actually of vegetable origin, but psychologically, it helps me to feel as though I'm having "meat" plus veggies. It also tricks my mind into thinking that I am splurging on something pleasurable from my pre-BLITZ diet days.

Countries like Japan, where plenty of soy is consumed, have very low rates of heart disease. Researchers at the University of Kentucky found that a soy-rich diet produced 12.9 percent and 9.3 percent declines in LDL and total cholesterol, respectively.

Eliminate high-fat cheese from your diet. When you're seeking substitutes, try the very-low-fat and non-fat cheeses. Taste a few brands, looking for those that please your palate. Since it's the butterfat that gives whole-milk cheese its texture and flavor, you may have to search rather aggressively for an alternative (particularly a non-fat one) your taste buds will find appetizing. But through trial and error, you will probably find one that's Near 'Nuff to the high-fat cheeses to melt on your veggie burger or sprinkle on pasta. Hang in there; I found several. But I can't guarantee that your tastebuds will pick the same ones I did.

Prepare your scrambled eggs or pancakes using egg whites or Egg Beaters, because all the cholesterol in eggs is in their yolks. Rainey and I have also begun to rely on a special oat-bran pancake and waffle mix marketed by Arrowhead Mills. When we're ready to serve the pancakes, we've forsaken traditional butter, substituting two products instead. One is a liquid: Fleischmann's Fat Free Low Calorie Spread. The other is a spray: I Can't Believe It's Not Butter!. Then we top them with maple-like syrup that's made with NutraSweet instead of sugar (look for it in the "syrup" or diabetic foods" sections of the supermarket). This alternative to traditional pancakes tastes just like the real thing, plus it is a delicious way to add oat bran fiber to my diet.

Complement your breakfast meal with low-fat side orders. On Sunday mornings, we have "bacon" with our pancakes. In our search for Near 'Nuff foods, we discovered Morning Star Farms Breakfast Strips imitation "bacon" strips, which look like bacon and have a taste that's about a 98% carbon copy of the real thing. Perhaps most interestingly, when you place these low-fat substitutes in a nonstick pan and heat them on the stove, they emit an aroma strikingly similar to bacon -- and they cook like bacon, too, complete with the little white bubbles that form on it. It's very clever and delicious. There's also a "sausage" version available.

In the kitchen, try doing something creative with portabella mushrooms. Rainey has discovered that when portabellas are grilled, they are a Near 'Nuff substitute for steak. She has even made sandwiches using a variety of vegetables, including grilled portabellas. (See the "Mediterranean Mouthfuls" recipe on page 98.)

Satisfy your sweet tooth with delicious alternatives. Desserts don't have to taste like cardboard to be "nutritionally correct." There are plenty of Near 'Nuff choices to keep your appetite satisfied. For example, I find the taste of Haagen-Dazs chocolate frozen sorbet (a frozen treat on a stick) just as tantalizing as chocolate ice cream -- without all the fat!

You'll find many other Near 'Nuff choices in other parts of this monograph. In your own hunt for these alternatives, let your taste buds be your guide. It isn't enough for Rainey and me to find foods whose labels indicate they're low in fat; we go further and reject any that don't pass our personal taste test, too. As a result, we often have to try several brands before finding those that have the taste and texture of the real thing. In your own search, don't give up if the first product or two you try aren't on target; once you locate the winners, you'll never feel deprived making the switch to healthier eating.

NEAR 'NUFF™ FOODS

"Near 'Nuff"™ is a term we invented (and trademarked) to describe foods with lower levels of fat, cholesterol, sugar, and/or sodium than other better-known food items with undesirably high levels of these ingredients. Near 'Nuff foods are <u>near enough</u> (hence the name) in appearance, texture, taste and smell to be an acceptable substitute for these unhealthy food items.

Why use Near 'Nuffs?

It can be a little scary to start a new food program. You may ask yourself, "Will I feel deprived, hungry or discouraged without all my old comfort foods?" If you anticipate you might face this challenge, then give yourself a "crutch" to help adjust to new food tastes and more healthy combinations by using my list of Near 'Nuffs. This is just a starting place with a list of the foods that helped me, but by no means is it a complete list of all the substitutes that may currently be available. I invite you to research your local grocery stores, find new ones and report to me with your own discoveries. And take heart, the food manufacturers are striving to meet the demands of healthy eaters by continually adding new and better products to the grocery shelves. You're sure to find something to your liking. (To help you stay up to date you can subscribe to our newsletter, described at the back of this book.)

NEAR 'NUFF DAIRY PRODUCTS

Butter Substitutes:
Nabisco's Fleischmann's Fat Free Low Calorie Spread (5 calories)
I Can't Believe It's Not Butter! spray (0 calories)
Cream or Milk Substitutes:
Skim Milk
Carnation Light Evaporated Milk
Pet Evaporated Skimmed Milk
Sour Cream Substitutes:
Naturally Yours Fat Free Sour Cream
Land of Lakes Fat Free Sour Cream
Breakstone Fat Free Sour Cream
Yogurt
Dannon No Fat Plain Yogurt
Dannon Light (Fat Free flavored yogurt)
Cottage Cheese
Light 'N Lively Fat Free Cottage Cheese
Breakstone Fat Free Cottage Cheese

Other Cheese Substitutes:

Philadelphia Fat Free Cream Cheese

Cheese Slices (fat free or low fat):
Kraft Fat Free American, Swiss, and Sharp Cheddar
formagg®
Alpine Lace
Healthy Choice American
Borden Swiss, Sharp Cheddar

Grated fat free cheeses:

formagg® Non-Fat Grated Cheddar Cheese

Alpine Lace

Healthy Choice Pizza cheese, mozzarella, and cheddar

Sargento Light (small amount of fat, use sparingly)

Polly-O Mozzarella

Polly-O Ricotta

Egg Substitutes:

Egg whites from the whole egg -- discard the yellow yolk

Second Nature Eggs

Better n Eggs

Egg Beaters

NEAR 'NUFF SALAD DRESSINGS

Kraft Fat Free Salad Dressings: Ranch, Blue Cheese, and Honey Dijon

Wishbone Fat Free Italian

Alessi Balsamic Vinegar (aged)

Alessi White Balsamic Vinegar

Progresso Red wine vinegar

First Cold Pressed Virgin Olive Oil (many brands)

NEAR 'NUFF CEREALS

Cold Cereals:

Health Valley: Real Oat Bran, 100% Natural Bran, Healthy Fiber Flakes

Ralston Multi-Bran Chex

Nabisco Shredded Wheat 'Bran

Kellogg's All-Bran Extra Fiber, Crispix, Common Sense Oat Bran

Post Grape Nuts

Hot Cereals:

Quaker Oat Bran

Quaker Oatmeal

NEAR 'NUFF BEVERAGES

Crystal Light Mixes: Lemonade, Iced Tea, Fruit Punch (Sugar-free and non-carbonated)

Canfield Chocolate Sodas (Plain or see Recipe section for Leslie's After Dinner Chocolate Soda "milkshake" Concoction)

Low-Sodium V-8 Juice

Alcohol Free Beer: O'Doul's

Alcohol Free Wine: Ariel or Sutter Home Fre'

Herbal Teas

NEAR 'NUFF SOUPS

Campbell's Healthy Request Cream of Chicken, Tomato, Cream of Mushroom

Campbell's Reduced fat Cream of Celery

Swanson's fat free chicken broth

Swanson's fat free vegetable broth

Baxter's fat free onion soup

NEAR 'NUFF MEATS OR MEAT-LIKE ITEMS

Morning Star Breakfast Strips (fake bacon)

Eggplant (nature's meat)

Portabella Mushrooms (nature's hamburger)-large, round, flat mushroom cap that fits perfectly in a hamburger bun

Gimme Lean Meatless "ground beef"

Oscar Mayer Fat Free Bologna, Hot Dogs, Franks, Ham

Vegetarian hot dogs:

Garden dogs

Natural Touch

Soy Boy

Some Veggie Burger choices:

Morning Star Farms Meatless Garden Vege Patties

Grillers

Veggie Burgers (plain and black bean),

Better 'n Burgers

The Better Burger (veggie burger)

Garden Mexi Vegetarian Burgers

The Original Garden Burger

Green Giant Veggie Burger products

NEAR 'NUFF SIDE DISHES

B&M 99% Fat Free Brick Oven Baked Beans (vegetarian)

NEAR 'NUFF CONDIMENTS, FLAVORS & SEASONINGS

Nayonaise (substitute for mayonnaise; available at health food stores: 3g fat and 35 calories per tablespoon)

mustard

Carey's No-Fat Sugar Free Syrup (for pancakes)

Morton Salt Substitute

NoSalt

Vegit All Purpose Seasoning

Mrs. Dash (salt free) Fine Ground Herbs & Spices

Mrs. Dash Extra Spicy (salt free)

Spike (salt free seasoning)

McCormick All Purpose Table Shake

McCormick All Purpose Seasoning

McCormick Lemon Pepper

McCormick Garlic and Herb

lemon herb seasoning

Creole seasoning

chicken and beef bouillon cubes or powder

Goya Ham Seasoning (powder)

First cold pressed virgin olive oil
Pam Spray flavors: Olive Oil, Butter, Olive Oil & Garlic
boiled, mashed potatoes -- used as a thickening agent
Maggi seasoning
Worcestershire Sauce
Tabasco
Tiger Sauce
Pickapeppa Sauce
fresh herbs: cilantro, basil, dill, chives, rosemary, arugula, mint, etc.
Horseradish

NEAR 'NUFF DESSERTS AND SNACKS

Jell-O Fat Free Sugar Free Pudding: chocolate and vanilla
Haagen Dazs Chocolate Sorbet Bar (frozen)
fat free pretzels
air popped popcorn
Guiltless Gourmet fat free tortilla chips
Guiltless Gourmet fat free dips
pre-made or home made fat-free salsa
Mission Fat Free Tortilla Chips
Louise's Fat Free Potato Chips
Crackers
- Health Valley Fat Free Crackers (Whole Wheat, Pizza, Fire Crackers)
- Nabisco Harvest Crisps: 5-Grain, Garden Vegetable, Italian Herb

■ ■ ■ ■

Chapter 4

Breakfast And Beyond: More Super-Strategies For Cholesterol Busting

Choose breakfast foods designed for cholesterol reduction

Believe it or not, before I changed my diet as part of the 30-Day BLITZ, my favorite breakfast cereal was Kellogg's Frosted Flakes -- coated with sugar and bathed in skim milk. (I had fooled myself that it was healthy, not realizing it was only a half-hearted measure.) Frankly, I thought this was fabulous, and saw no reason to eat any other kind of cereal -- until I got the ominous message about my high blood cholesterol and triglyceride levels.

That's when my search for a cereal rich in soluble fiber began. If I chose it carefully, I reasoned, my breakfast cereal would be a way to start most days with cholesterol-lowering clout. So I began reading labels, and found a lot of ready-to-eat cereals that were bursting with fiber and were low-fat or fat-free, such as Kellogg's All Bran with Extra Fiber or Post 100% Bran. Frankly, a complete list of these healthy breakfast foods could be almost as lengthy as a supermarket aisle. Fiber, however, can be of many types: wheat bran, corn bran, etc. Be sure to put at the top of your list cereals high in oat bran, and fill in with secondary brans as you wish.

FIBER-RICH CEREALS I USE IN MY BLITZ

Cold Cereal:	Fiber Content per serving:
Kellogg's All Bran w/Extra Fiber	15 grams
Post 100% Bran	8 grams
Nabisco Shredded Wheat N Bran	8 grams
Ralston Foods Multi-Bran Chex	7 grams
Health Valley 100% Natural Bran Cereal w/Apples and Cinnamon	7 grams
Post Shredded Wheat Spoon Size (no salt or sugar added)	5 grams
Post GrapeNuts	5 grams
Health Valley Real Oat Bran Cereal	5 grams
Kellogg's Common Sense Oat Bran	4 grams
Health Valley Organic Healthy Fiber Multi-Grain Flakes	4 grams
Kellogg's Crispix *	1 gram
Hot Cereal:	
Quaker Oat Bran Hot Cereal	6 grams
Quaker Oatmeal Hot Cereal	6 grams

** Rainey uses a small amount of Crispix to add good "crunch" to the other high fiber cereals that need a crispy boost to suit her taste.*

My suggestion is that you spend some time reading labels, and finding at least one or two brands of high-fiber, low-fat cereal that appeal to you (too many cereals may look health-promoting, only to be packed with too much fat, sugar, and sodium, and too many calories). Also, if you're not taking vitamins and mineral supplements (as I do), look for cereals that have been fortified with vitamins (like A, D and B_{12}) and minerals (including calcium), sometimes at levels that meet the RDAs for these nutrients.

Rainey and I ended up buying about a dozen fiber-rich cereals; but I must admit that my taste buds initially didn't absolutely fall in love with any of them. Even so, I've found a way to make breakfast interesting. Here's how: I place the boxes of cereals in a row in front of me on the kitchen counter, and to avoid boredom, create my own multilayer "cake" with them -- pouring a little of this and that into a bowl, and making something resembling a cereal "sandwich." Then I wet it down with skim milk, and sprinkle half a teaspoon of sugar over the surface to sweeten it a little (an envelope of artificial sweetener can be used instead). Yes, I know some of you "purists" will say I'm "contaminating" the cereal, but it's a small price to pay for getting rid of fat and gaining a lot of fiber. The cereal can also be topped with a few slices of fruit (bananas, strawberries, blueberries - a very high fiber fruit).

This cereal blend has a variety of flavors, consistencies and crunchiness, and it's actually very interesting and a little different every day. Sure, it's not my old friend, Frosted Flakes, but I now look forward to indulging in my breakfast concoction each morning, and certainly don't feel that I'm depriving myself. Funny enough, too, that after months of this my taste buds changed, and now sugar-rich cereals are too sweet for me.

Now, what if you'd prefer to start your day with something warm in your stomach? There are plenty of hot cereals that are very low in fat (Cream of Wheat and farina, for example). I personally eat a lot of Quaker oatmeal and oat bran -- or a mixture of these. But again, read labels carefully, and don't spoil a healthy cereal by bathing it in butter or cream when serving.

For a real change of pace from the snap, crackle and pop of many of your mornings, try some of the alternatives mentioned earlier in this monograph --

such as pancakes made with Arrowhead Mills oat bran pancake and waffle mix (we prepare them with egg whites and a little canola oil). The mix has a nutty, almost vanilla, taste that we find truly delicious. We serve the pancakes with a side order of "bacon" -- Morning Star Farms Breakfast Strips imitation bacon.

And we like our pancakes slathered with "butter" and "syrup". So, in our program we use two butter substitutes. Both have zero fat. One is a spray: I Can't Believe It's Not Butter!. The other is a liquid in a squeeze bottle: Fleischmann's Fat Free Low Calorie Spread. Then we pour on maple-flavor, reduced-sugar or artificially sweetened syrup. All these goodies can be found in your supermarket.

We also occasionally have an oat-bran muffin with breakfast. You've probably heard a lot about oat bran in recent years -- in the 1980s, a popular book turned the rush for oat bran into a real stampede, with people clinging (erroneously) to the hope that oats were a "miracle food" that would magically grant them good health. There were even several studies confirming that oats could help bring your cholesterol under control (although it certainly never possessed "magic bullet" properties). For example, in research at the University of Kentucky, men with high cholesterol added either oat bran or wheat bran to their diets by incorporating them into muffins or cereal. Three weeks later, those eating oat bran experienced nearly a 13 percent decline in their total cholesterol; by contrast, there were no decreases in the cholesterol counts of the wheat-bran group.

So what's the real scoop on oats? Well, oat bran does have plenty of soluble fiber, which probably is responsible for its genuine cholesterol-cutting capacity. But beware: While you may be able to buy oat-bran muffins in a neighborhood bakery, you never really know what you're getting unless you specifically ask the baker or read labels carefully; many commercial muffins may be brimming with oat bran, but they could be filled with fat as well.

What's the alternative? Frankly, I haven't found commercial oat bran muffins as tasty -- and as low in fat -- as those that Rainey bakes. Her recipe for these muffins is on page 92 of this monograph. We don't eat oat-bran muffins every day, but consider them a treat; after you taste them, you'll see why.

And last but not least, the all-time breakfast favorite -- eggs. Aren't you shocked we use them? After all, a whole egg contains a lot of fat and cholesterol. Well, here's the important secret for you to know and follow: we use only the egg white. The white contains all protein and no fat. Discard the yolks; it contains all the cholesterol and is 82% fat. (Rainey likes her eggs to look yellow, so sometimes she prefers to use Egg Beaters or a similar egg substitute.)

Here's another bonus. The white of a large egg contains only 12 calories, so you can have a four-egg-white omelette and ingest only 48 calories! Try mixing a few chopped veggies, like mushrooms, onions, or tomatoes for extra flavor.

So overall, you can say eggs have gotten a "bad rap". Take advantage of the fact that Nature compartmentalized the egg into a "good" part (the white) and a "bad" part (the yellow yolk). Use just the whites and enjoy yourself!

Choose healthy snacks such as fresh fruit and raw veggies

Mention the word "snacks," and many people think immediately of high-fat cookies, cheese crackers, and potato chips. But to keep your goal of cholesterol-cutting in the forefront, you need to shift your attention to snacks rich in complex carbohydrates and/or fiber. That means taking a bite out of non-fat fresh fruit and raw vegetables, which are brimming with fiber (soluble and insoluble) and antioxidants (such as vitamin C and beta-carotene) -- both of which are very heart-friendly.

Carrot sticks and celery stalks, for example, may not sound particularly exciting at first glance (they've probably been part of most weight-loss programs you've tried). But research suggests that both have unique cholesterol-crunching properties. Carrots are laden with a type of fiber called calcium pectate, which can play a crucial role in moving cholesterol out of the body. In fact, consider this finding: A small Scottish study concluded that after eating just two carrots a day for three weeks, a group of men experienced an average decline of 11 percent in their total cholesterol counts.

But what if carrots sound a little too dull for your own taste? There are many ways to fight for flavor -- for example, by dipping raw carrots in fat-free sour cream or yogurt. A little chopped dill or other herbs can also provide vegetables with a flavor boost without adding fat. Rainey loves salsa as a taste-enhancer (but it isn't my personal favorite). I prefer non-fat ranch dressing or another type of non-fat salad dressing, pouring a tiny puddle of it on my plate, and then dipping the tip of each vegetable piece in it. Even though the dressing is fat-free, I don't load up the entire curvature of a celery stalk with it, but use it only sparingly. Remember, carrots and celery aren't porous, and thus they don't absorb much dressing; just a modest surface coating gives them all the extra taste I want.

As for fruits, most provide their own sweet flavor, and I've already described topping your cereal with strawberries, bananas or blueberries. For snacking, why not stir some sliced apples or peaches into a dish of non-fat yogurt? And from time to time, spice up your diet by choosing less frequently-eaten fruits like kiwi, guava, and raspberries.

Popcorn is a great snack, too, as long as it's air-popped and not smothered in butter, margarine or salt (limiting salt is particularly important if you're also trying to put a cap on your blood pressure). A typical serving of popcorn (three cups) contains about four grams of fiber, approximately half of which is soluble. After it's popped, to give it a butter flavor, "spritz" it with I Can't Believe It's Not Butter! spray. Instead of salting it with table salt, use a salt substitute.

A distant second choice is microwavable popcorn, but make sure it's "light". Read the nutritional label carefully to ensure that it's truly low in saturated fat. If you choose to prepare the corn in oil rather than air-popping or microwave, my best kernel of advice is to avoid saturated fatty oils (like coconut oil), opting instead for one rich in monounsaturated fat like canola oil. However, even if the oil is monounsaturated, it's still fat. Because popcorn is porous, oil will seep into the popcorn and end up in your body -- so the less of it you use, the better. However, to ensure great results from the 30-Day BLITZ, avoid using oil on popcorn until you are on your maintenance program. Remember, the 30 days is your chance to do everything possible to enhance your chance of lowering your cholesterol.

What about other snack selections? Pretzels are a better alternative than potato chips. Many well-known brands now offer non-fat varieties. Rice cakes are preferable to chocolate-chip cookies. Even a few almonds or other nuts are OK -- as long as you don't ignore the word "few." Most nuts have lots of monounsaturated fat (about two-thirds of the total fat in almonds is monos). In a study at the Health Research and Studies Center in California, volunteers added three ounces of raw almonds to their diets, and began using almond oil in place of other vegetable oils. After three weeks, even though their overall fat intake had increased, their total cholesterol fell an average of 20 points (from 235 to 215).

Still, I can't overemphasize that you shouldn't go nuts over almonds (or any other snack that contains fat). If almonds are something you enjoy, have a handful (or sprinkle a few sliced ones over your salad or cereal); if you eat them in moderation in place of other saturated-fat-congested foods, they can become a team player in your cholesterol-busting game plan.

Choose healthier desserts

You don't have to place your sweet tooth into retirement to climb aboard the BLITZ bandwagon. Rainey and I have discovered delightful desserts just by sleuthing the freezer sections of supermarkets. For instance, we have found many varieties of non-fat, low-calorie ice cream made with fat substitutes and NutraSweet. Read labels carefully, however, since some "diet" ice cream products are high in both carbohydrates and fats. Also, be willing to conduct a few taste tests until you find the perfect products for you.

Here are just a few of the specific desserts that Rainey and I have discovered: As I mentioned earlier, I've become a big fan of Haagen-Dazs fat-free chocolate frozen sorbet-on-a-stick. It's just as taste-tantalizing as high-fat chocolate ice cream. Look for sherbet, too; a cup contains just 2 to 4 grams of fat -- about one-sixth or less of the amount of fat in the same serving size of premium ice cream. Rainey found non-fat Jell-O pudding, made with NutraSweet, in flavors like chocolate fudge. It's really fabulous, especially when you realize there's no fat or sugar.

We make our own tasty drinks with dietetic chocolate soda, although again, we had to search for just the right brand. We found only one -- Canfield's -- that's "chocolaty" enough for us; we pour it into a little skim milk, and it foams up like a chocolate soda with cream (See the recipe on page 112).

So when it comes to desserts, we really haven't suffered. Angel food cake is made with egg whites, and has only trace levels of fat. Frozen juice bars -- made of 100% fruit -- are good choices, too. And don't forget fresh fruit as a dessert -- such as sliced strawberries rather than a piece of strawberry pie (which tends to be bursting with saturated fat and calories). However, it's important to remember to have modest portions. I used to fool myself by eating huge amounts of non-fat deserts that happened to be loaded with calories. But, many of those extra carbohydrate calories are stored by your body as fat.

By the way, I don't eat cakes and cookies anymore (even non-fat ones), but perhaps you'll find some low-fat varieties that you enjoy. Before I began the 30-Day BLITZ, I ate a lot of Entenmann's no-fat pastries, particularly the pound cake -- but was oblivious to the fact that while it was truly fat-free, it was very high in calories. That's troubling news, especially since I would sometimes eat half of a one-pound cake at a sitting! So read labels; you might be surprised at what you find.

■ ■ ■ ■

Chapter 5

The Little Things That Count

Choose your condiments carefully

There's no quicker way to sabotage a low-fat dish than to soak it in taste enhancers like a cream sauce or melted butter. You can even "ruin" a perfectly healthy slice of wheat toast by overloading it with cream cheese, butter or margarine.

After all of the negative talk about butter in recent years, perhaps you joined the switch to margarine, and believed that you had finally found a health savior in all this "margarine mania." However, dietitians now know that even traditional margarine isn't problem-free: Margarine is still 100% fat (even though it's mostly unsaturated). Perhaps even more unsettling, it contains a kind of "bad fat" called trans-fatty acids, which can actually increase your LDL cholesterol level (the "bad" cholesterol).

For these reasons, I believe you're better off putting a little jam or jelly on your toast (Though they have calories from sugar, they're completely free of saturated fat.), or choosing the butter substitutes that Rainey and I have found. As I've mentioned, I'm not a big fan of deprivation, particularly when my taste buds are concerned. But we've discovered that you can have wonderfully tasty meals and still successfully win the cholesterol wars using products such as liquid Fleischmann's Fat Free Low Calorie Spread, and the spray, I Can't Believe It's Not Butter! -- both of which are wonderful alternatives to traditional butter and margarine. But when you're shopping for Fleischmann's, pass up their solid margarine, even the low-fat version, and look for their liquid "butter" in the squeeze bottle, which is completely fat-free. It smells and tastes like butter, and is unbelievably good.

When shopping for I Can't Believe It's Not Butter!, avoid the tub (it still has calories and fat, however low). Look for the spray instead, which has absolutely no fat and few calories. Sure, in a blindfold test, maybe chefs from Paris could detect the difference between both of these products and the real thing; but to my ordinary taste buds, if I spray or dab these imitators on my finger and take a lick, it's just about the same as tasting actual butter.

We lightly spread the Fleischmann's liquid butter on toast, and put it atop baked potatoes. We use the spray to give a butter-like flavor to vegetables and popcorn. And frankly, we can't tell the difference. Sometimes when I have a butter craving I use *both!* (Yet still there's no fat.)

By the way, there are plenty of other ways to spice up that baked potato. In addition to non-fat butter, Rainey and I use a little non-fat sour cream, and then we sprinkle on a few Bacos (imitation bacon bits) made from soybean protein, plus some fat-free grated cheese. And it tastes great. In fact, as we're enjoying these "loaded" baked potatoes, I sometimes say to Rainey, "Can you believe we're eating entirely non-fat?!"

What about salad dressings? Of course, supermarket shelves have an array of non-fat selections. But if you use oil (for dressings or for cooking), use it sparingly, and remember to rely on choices like cold-pressed virgin olive or canola oil (as the accompanying chart shows, they are primarily monounsaturated). Some people enjoy the Mediterranean tradition of dipping their bread in a little bit of olive oil, or brushing a small amount of the oil on corn on the cob (in place of butter), or sauteing their fish in it.

When consumed in moderation, olive oil has some amazing health-promoting qualities. In a number of studies, it has been shown to not only reduce your bad (LDL) cholesterol, but also to increase your good (HDL) cholesterol. Researchers in Spain, for instance, put women on a diet high in saturated fat (including butter and palm oil) for a month, and then switched them to a diet containing olive oil for six weeks, followed by a third diet rich in a polyunsaturated fat (sunflower oil), again for six weeks. Both total and LDL cholesterol decreased on the latter two diets; but only the olive oil diet also increased HDL ("good") cholesterol. On the sunflower oil regimen, HDL fell.

Again, however, moderation is the key: Olive oil is 100 percent fat, and it's bursting with calories, too. If you overdo it, not only can your coronary arteries suffer, but you may gain weight, too, which can escalate your chances of developing high blood pressure and an array of other obesity-related diseases. Fortunately, because olive oil has such a robust flavor, just a little is all you really need.

WHAT FATS ARE YOU CONSUMING?

All vegetable oils are a mix of saturated, polyunsaturated and monounsaturated fats. In each type of oil, one type of fat tends to dominate, and thus nutritionists place it in that particular category. Thus, because olive and canola oils are high in monounsaturated fats, they are considered monounsaturated; while safflower and sunflower oils are categorized as polyunsaturated. Remember, no fats should be consumed in excess, but when you must use them, you are still better off choosing those high in monounsaturated or polyunsaturated fats -- instead of saturated ones.

Table 3

Fats in Common Cooking Oils

	Mono	**Poly**	**Saturated**
Canola	62%	32%	6%
Safflower	12%	75%	9%
Sunflower	20%	66%	10%
Corn	24%	59%	13%
Olive	72%	9%	14%
Soybean	23%	59%	14%
Peanut	46%	32%	17%
Sesame seed	40%	40%	18%
Palm kernel	10%	2%	80%
Coconut	6%	2%	87%

Notes:

1. Arranged from best down to worst, according to content of saturated fat.

2. The percentages of nearly all of these oils do not total 100 percent because each includes small amounts of fat-like constituents.

Avoid alcohol

Before I began the BLITZ, I would "unwind" from the day's stresses by pouring myself a few glasses of wine (or a half or whole bottle!) with dinner. However, when Rainey and I started researching the effect of diet upon heart-disease risk factors, we read studies showing that alcohol could sometimes raise triglyceride levels, in part by increasing the production of triglycerides in the liver (remember, my triglycerides were very high, which was just as alarming to me and my doctor as my elevated cholesterol). We also learned that some hypertensives have an alcohol sensitivity -- they can have a couple drinks, and their blood pressure climbs in the process. So because my sights were set on reducing both my high blood pressure and triglycerides, I decided to swear off alcohol -- at least during my 30 day BLITZ.

Of course, I'm also not blind to the evidence that alcohol in moderation -- such as a glass or two of wine, particularly red wine, a day -- may raise your HDL cholesterol level (by an average of about 5 points, according to one study), and thus reduce your risk of a heart attack. No one knows for certain precisely how alcohol exerts this protective effect, but it does seem to be a genuine benefit. On balance, however, I decided that alcohol had more potential to do harm than good in my own life; besides, I was also eliminating several hundred calories per day by omitting it from my 30 day BLITZ.

I agree with those doctors who tell patients, "If you don't already drink, the possible benefits of alcohol upon your HDL is no reason to start." That's particularly true in light of the devastating evils of excessive alcohol consumption (alcoholism, liver disease, cancer, heart rhythm irregularities, and birth defects). A study at Kaiser-Permanente Medical Center in Oakland, California, concluded that while one or two drinks a day can cut the risk of heart-disease deaths by 30 percent, heavy drinkers (six or more drinks a day) actually had a greater likelihood of death from noncardiovascular conditions, compared to non-drinkers.

Even though I decided to cut alcohol out of my life, I was determined to avoid sacrificing the sensation and pleasure of "having a drink". And so I looked for and found a number of alternatives, which you might also decide to try as well. For instance, there are many non-alcoholic "substitute" (or as we prefer to say, Near

'Nuff) drinks available -- one of my favorites is O'Doul's Non-Alcoholic Beer. Rainey and I have also found two brands of nationally distributed non-alcoholic wines that are particularly good: Ariel and Sutter Home Fre´.

I've also created my own non-alcoholic version of a Bloody Mary -- a spicy, low-sodium V-8 juice drink. As well as being very tasty, its fiery spiciness is designed to send my endorphin levels (the brain's own "feel-good" chemicals) soaring. It also tricks my taste buds into believing that I'm really drinking a cocktail, subduing any lingering craving for an alcoholic drink. Just like a real Bloody Mary, my concoction is a mix of Worcestershire sauce, Tabasco, and lemon juice. But instead of tomato juice, which is very high in sodium, I use "low-sodium" V-8 juice, which is rich in potassium. For this particular "cocktail," V-8 juice also simply tastes much better than "low sodium" tomato juice. I blend the ingredients, and serve it with a stalk of celery -- its spiciness just about produces beads of sweat on my brow.

When searching for other non-alcoholic drinks, a simple glass of grape juice is worth considering. Researchers believe that some of the benefits of red wine may lay in its grape pulp; it gives red wine its color, and also contains substances such as flavonoids. These are antioxidant compounds that protect the heart by minimizing the "stickiness" of blood platelets -- thus reducing the likelihood of blood clots and heart attacks. A study at the University of Wisconsin found that Welch's natural purple grape juice had anti-clotting qualities similar to those of red wine.

Every one of these alternatives gives me the social ritual of "having a drink," without the downside associated with alcohol. The non-alcoholic wines I've mentioned are even marketed in wine-like bottles, complete with a cork. They're the color of wine, and have almost the same taste. And when you go through the ritual of screwing out the cork and pouring the liquid into a wine glass and sipping it, it's about as close as you can get to the real thing without getting tipsy.

Take advantage of other cardiovascular protectors

Some of the BLITZ's recommended dietary changes (like eliminating alcohol) are designed primarily to reduce blood pressure and affect triglycerides.

However, because of the interconnectedness of the entire cardiovascular system, I theorize that these same strategies may have a beneficial effect upon blood cholesterol levels as well -- directly or indirectly. Thus, while the primary "thermometer" I'm eventually trying to affect is my cholesterol count, I nudge it from as many directions as possible, including those whose primary effect may be upon other components of the cardiovascular network.

For instance, because I keep a close eye on my blood pressure, I remove as much salt as possible from my diet (about 25 percent of hypertensives are salt-sensitive, meaning that the more sodium they consume, the higher their blood pressure will climb). One study found that short-term, moderate sodium restriction can reduce systolic blood pressure by an average of about 5 millimeters, and diastolic pressure by 2.6 millimeters.

In place of table salt, Rainey and I often use herb preparations and pepper powders that are available in the seasoning sections of most markets. Some come in shaker bottles, so you can sprinkle them just as you would salt.

At the same time that I've reined in my ordinary table salt consumption, I've added potassium salt to my diet. It doesn't have exactly the same taste as sodium chloride (table salt), but it's similar, and is sold in the markets as a salt substitute. Some research suggests that as you increase your intake of potassium (as a substitute for salt), you can add an additional element of protection against high blood pressure. And, believe me, while I love the taste of table salt, I don't mind this switch at all. A special note of caution: Potassium can cause negative side effects for certain groups, such as people with impaired kidney function, as well as those taking "potassium sparing" diuretics. Excess potassium can build up to a dangerous level under these circumstances. Check with your doctor before adding it to your diet.

By the way, many food manufacturers sneak hidden sodium salt into processed foods -- such as canned soups, ketchup and snacks -- so read all labels and avoid products that are sodium or salt-intensive.

I also make an effort to eat plenty of foods high in magnesium -- such as leafy green vegetables, whole grain cereals, bananas, and legumes. Magnesium has particularly powerful effects on insulin regulation, and if your consumption of magnesium is low, the shortage may contribute to heart disease, high blood pressure and Type II diabetes. Some research has suggested that vegetarians tend to have lower blood pressure because of their higher intakes of magnesium, potassium and calcium, along with a reduction in dietary fat and sodium.

FOODS HIGH IN POTASSIUM AND LOW IN SODIUM

Black Beans	Asparagus
Pecans[1]	Peas
Dates[2]	Corn
Acorn Squash	Apple
Banana[3]	**Avocado**[3]
Lima Beans[3]	Eggplant
Soy Beans	Oatmeal
Orange	Zucchini
Orange Juice	**Potato**[3]
Strawberries	Blueberries

Notes:
1. Eat only a modest amount, as pecans are very high in fat.
2. Eat only a modest amount, as dates are very high in natural sugar.
3. These foods are also high in magnesium.

FOODS HIGH IN MAGNESIUM

Bananas*	Kidney Beans
Blackeyed Peas	**Lima Beans***
Buckwheat Flour	**Avocado***
Whole Wheat Flour	**Potato***

Notes:
* These foods are also high in potassium.

Read food labels

Take a moment to look at the Nutrition Facts labels on most food products. You might be surprised at just how packed with information they are nowadays -- and how helpful they can be in making the transition toward lower-fat eating. Since the day Rainey and I began creating the 30-Day BLITZ, these labels have guided us in shopping for low-fat foods, and frankly, we've been delighted at just how many heart-healthy choices are available today.

The Nutrition Facts label lists a product's total calories, total fat, saturated fat, cholesterol, sodium, dietary fiber, vitamins, minerals, and much more. But for the purposes of making low-fat choices, pay particular attention to the amount of "total fat" (some experts believe you should aim toward keeping your overall daily fat intake at less than 75 grams a day), "saturated fat" (less than 25 grams a day), "calories" and "calories from fat." If, as in the sample label shown here, a food's serving size has 90 calories, and 30 of those calories are from fat, then 33 percent of its calories are fat calories; that's really too high for a low-fat eating plan, where your overall fat intake should be less than 30 percent (I personally aim for much lower than that 30 percent target).

Also, when choosing between brands, select the one with the highest fiber content. Unfortunately, the Nutrition Facts label does not break fiber down into soluble and insoluble, but the higher the overall fiber content, the better.

Finally, always look carefully at the "serving size." The manufacturer may have listed a serving size a lot smaller than the amount you choose to eat, and that can get you into trouble. For example, if a package of cookies lists the serving size as two cookies -- but you decide to eat four or six instead -- you'll be consuming two to three times as much fat and calories as you might think by glancing at the label. (I've also noticed how a few companies with high calorie or high fat products try to get around this problem by listing serving sizes that are unrealistically low -- for example, "one cookie". This lets them print lower calories and fat "per serving". Don't be fooled.)

Nutrition Facts

Serving Size 1/2 cup (30g/1.1 oz.)
Servings per Container 11

Amount Per Serving	**Cereal**	**Cereal with ½ Cup Vitamins A & D Skim Milk**
Calories	50	90
Calories from Fat	10	10
	% Daily Value **	
Total Fat 1.0g*	**2** %	**2** %
Saturated Fat 0g	**0** %	**0** %
Cholesterol 0mg	**0** %	**0** %
Sodium 150mg	**6** %	**9** %
Potassium 350mg	**10** %	**16** %
Total Carbohydrate 22g	**7** %	**9** %
Dietary Fiber 15g	**60** %	**60** %
Sugars 0g		
Other Carbohydrate 7g		
Protein 4g		
Vitamin A	15 %	20 %
Vitamin C	25 %	25 %
Calcium	10 %	25 %
Iron	25 %	25 %
Vitamin D	10 %	25 %
Thiamin	25 %	30 %
Riboflavin	25 %	35 %
Niacin	25 %	25 %
Vitamin B_6	25 %	25 %
Folate	25 %	25 %
Vitamin B_{12}	25 %	35 %
Phosphorus	30 %	40 %
Magnesium	30 %	35 %
Zinc	25 %	25 %
Copper	15 %	15 %

*Amount in cereal. One half cup skim milk contributes an additional 40 calories, 65mg sodium, 6g total carbohydrate (6g sugars), and 4g protein.
**Percent Daily Values are based on a 2,000 calorie diet. Your daily values may be higher or lower depending on your calorie needs.

	Calories	2,000	2,500
Total Fat	Less than	65g	80g
Sat. Fat	Less than	20g	25g
Cholesterol	Less than	300mg	300mg
Sodium	Less than	2,400mg	2,400mg
Potassium		3,500mg	3,500mg
Total Carbohydrate		300g	375g
Dietary Fiber		25g	30g

Calories per gram:
Fat 9 • Carbohydrate 4 • Protein 4

■ ■ ■ ■

Chapter 6

Pills And Powders: The Power Of Food Supplements

Just about anyone you ask will tell you that eating is one of life's real pleasures. No wonder many people cringe at the thought of relying on pills and powders to help reduce their cholesterol count. Although these food supplements are important, in the 30-Day BLITZ they are primarily an adjunct to the powerful (and delicious) dietary program that I've already described (and that you'll find in the recipe section later). Again, in creating the BLITZ, Rainey and I decided to attack the cholesterol problem from as many directions as possible all at once, and supplements are just one of the weapons with which we can promote our own good health.

Here are the supplements that I take every day; I recommend that you consider adding them to your own cholesterol-busting plan:

Apple pectin

One of the keys to conquering cholesterol with this program is to fill yourself up with fiber, particularly the soluble kind. Pectin is a widely available type of soluble fiber, and while it's present in most fruits and many veggies, apples are among the best places to find it. The average American eats more than a hundred apples a year, but why not pack even more pectin power into your diet with supplements?

Just consider the results of a study at Central Washington University. Researchers hypothesized that apple fiber could interfere with the absorption of cholesterol in the body, and in fact, found it could cut blood cholesterol levels by an average of 15 points in just six weeks. While that may not be a knockout punch, it will take a hefty bite out of your cholesterol count. And once pectin joins all the other items in your personal BLITZ, the overall impact can be dazzling.

I continue to fit as many real apples into my diet as possible, but I've also given apple pectin supplements an important supporting role. They're available in capsule or powder form and I take 2,000 mg. a day, divided among two capsules -- one in the morning and one at night. You can also buy apple pectin as a powder, and add a spoonful to various liquids. But if you're going to drink it, do so quickly. Otherwise, it "gels" and your glass of liquid could look like solid Jell-o after a few minutes of standing.

At breakfast, I also sometimes add powdered apple fiber (which contains soluble pectin plus fiber) to oat-bran pancake mix, low-fat oat-bran muffins, and even juices. Later in the day, I may use it as a thickener in sauces.

Oat bran capsules

You've already read about oat bran earlier in this monograph, and its reign as a media darling in the '80s. But while I generally don't believe in faddish shortcuts to good health, oat bran has proven itself in the world of cholesterol-cutting, and deserves a place in the BLITZ's winner's circle.

You can find oat bran in tablet or capsule form in many health-food stores. I take two capsules per meal (1,500 mg. per capsule) in my personal campaign against cholesterol. You can also try sprinkling oat bran powder on your breakfast cereal.

Guar gum powder

Don't be deceived because guar (pronounced gwar) gum hasn't achieved the celebrity status of oat bran; in fact, it's just as potent a weapon against high cholesterol counts. Ironically, however, you might not have even heard of guar gum. By its name alone, you might think it's something you chew or blow big bubbles with. But, in fact, it's a type of soluble fiber extracted from a legume called the cluster bean, which is grown primarily in Texas.

Until recently, if guar gum was used at all by food manufacturers, it was as a thickener in jams, jellies and other foods. But that was before millions of people became cholesterol-conscious. In recent years, researchers have found that guar gum can send cholesterol levels into a quick nosedive. In a study in India, for example, 20 volunteers consumed 15 grams of guar gum a day, adding it to flavored drinks and biscuit recipes. After six weeks, all the volunteers had experienced a cholesterol decline ranging from 10 to 27 percent.

Probably the simplest way to consume guar gum is in powder form, stirring it into a glass of chilled orange juice, lemonade, or other beverage. I consume four grams of guar gum (a heaping teaspoon) in the morning and another four grams in the evening. But as with pectin powder, stir it into the liquid of choice fast, and drink it quickly. Otherwise it starts to gel and if left standing too long, it can turn your glass of liquid into a gelatin-like substance. Also, drink plenty of liquid with it and don't let the dry powder get into your mouth; it can swell up as a gel and gag you.

Psyllium

The word "psyllium" (pronounced "silly-um") may not sound familiar. In fact, it might be one of the best kept secrets in the war on cholesterol.

Psyllium has actually been widely available for many years -- in both powder and wafer form -- but primarily it has been used as a compound to increase stool bulk and promote colon action, not as a cholesterol-buster. Scan the drug-store shelves, and when you pick up products like Metamucil, FiberCon and Correctol, you'll be looking at psyllium-rich formulations. Yes, they can relieve constipation, but at the same time, they'll also deplete your high cholesterol levels. Actually, they are not "laxatives", in that they do not directly irritate your colon and stimulate it to action. Rather, they absorb water and swell up, thus adding soft bulk to your stool. This increased bulkiness improves regularity.

Are you baffled at how a stool-bulking agent could possibly help keep your coronary arteries clear? Well, look no farther for an explanation than psyllium's fiber content. Psyllium comes from the seeds of certain plants (It doesn't sound too appetizing, does it?) and it is an excellent source of soluble fiber -- the kind that can draw cholesterol out of the bloodstream and the body.

Consider the findings of a study by researchers at the University of Cincinnati and Washington University. A group of individuals with cholesterol levels of 220 or higher added about 10 grams of psyllium (in orange-flavored Metamucil) to their daily diets. After two months, their LDL cholesterol levels plummeted about seven percent, whether they were eating a high-fat or a low-fat diet.

If you decide to take psyllium-rich laxative products -- either wafers or powders (the latter can be mixed into a glass of juice or water) -- follow the dosage instructions on the label. Also, keep in mind that there's still another way to add psyllium to your diet: Some ready-to-eat breakfast cereals, such as Bran Buds and FiberWise, contain psyllium; a small study at the University of Toronto found that people who ate Bran Buds (two-thirds of a cup a day) experienced LDL cholesterol declines of 11 percent after two weeks.

One important warning, however: If you're taking any prescription medication, talk to your doctor before adding psyllium to your personal BLITZ. Consumed in large amounts, psyllium can interfere with the body's absorption of certain drugs, including high blood pressure and heart medications.

So what's the bottom line on psyllium? I believe it can be a valuable component of a cholesterol-reducing program, although don't think of it as a primary source of dietary fiber. Yes, the laxative-like products I've mentioned are fiber-rich, but they won't provide you with other essential nutrients. So make sure you're also getting plenty of soluble fiber from foods (fruits, vegetables, grains) to help maintain a balanced diet. Also breakfast cereals like Bran Buds are a good source of both soluble and insoluble fiber, and thus you'll not only get a cholesterol-cutting clout from them, but the insoluble fiber will help protect you from colon cancer and other diseases of the digestive system.

Deodorized garlic capsules

When I began researching garlic, I discovered that it could do more than take points off my cholesterol count -- it could also help control high blood pressure. So for someone with my dual health problems of hypertension and high cholesterol levels, garlic had particular appeal.

Even so, I was faced with a dilemma: I knew I was hypersensitive to garlic, or at least I thought I was. When I would eat a garlic-rich meal, I'd have a lingering taste of garlic in my mouth for days (garlic consumption has been linked to heartburn, gas and occasionally allergic reactions). As impressive as the scientific research was favoring garlic, I couldn't imagine a way to comfortably add this odoriferous herb to my diet.

At the same time, however, I didn't feel I could completely ignore what garlic has to offer. Some investigators had hypothesized that a compound called allicin might be responsible for garlic's artery-clearing benefits, although there was some question whether allicin's benefits were lost in capsules. Nevertheless, as I continued to read the medical literature, I uncovered some studies showing that capsules could have a positive effect as well. A European report, for example, described 40 patients with high cholesterol levels, who were given either garlic powder (900 mg. a day) or a placebo. After 16 weeks, the total cholesterol of the garlic-powder group had declined 21 percent, while their triglycerides had taken a 24 percent plunge.

That's when I made the decision to incorporate *deodorized* garlic-powder capsules or tablets (which contain dehydrated or pulverized cloves) into the BLITZ -- an option you should consider if fresh garlic cloves leave you or your close-proximity acquaintances gasping for breath. Not only might they have benefits of their own, but they might also have a synergistic effect in combination with other components of our plan.

I take one deodorized garlic tablet a day -- typically, a 400-milligram tablet of Garlique; it's the equivalent of about a clove of garlic (a study at New York Medical College concluded that a dosage equal to one-half to one clove a day could cut into your cholesterol count). These tablets are a much better option than the garlic salt you might have spotted on the supermarket shelves; garlic salt is brimming with sodium -- which can wreak havoc on your blood pressure readings -- and there's no evidence that it can affect your cholesterol level in a positive manner.

By the way, while I thought I had good reason to keep fresh garlic cloves at arm's length, Rainey has eased both garlic and onions into our cooking, and strangely enough, my digestive system is getting used to them. But I'm still taking a garlic capsule every day as well, and it remains my primary source of garlic.

Wheat germ

If you're looking for an extra cholesterol fighter to sprinkle on your Bran Buds or other breakfast cereal, think wheat germ. Wheat germ is the "heart" of the wheat kernel, and at first glance, it might seem like a questionable addition to your anti-cholesterol game plan. After all, it has plenty of fat -- but the good news is that 60 percent of the fat is polyunsaturated, and another 17 percent is monounsaturated; so if you don't overdo it, wheat germ might actually become an ally in your personal cholesterol campaign.

What else does wheat germ have on its side? There's fiber (about two grams in a two-tablespoon scoop), plus vitamin E (which I'll discuss later for its ability to guard against heart disease). Wheat germ also has lots of magnesium, a mineral that's often deficient in people with high blood pressure and heart disease. Wheat germ has plenty of "phytosterols," too, which are compounds that some researchers believe can block the absorption of cholesterol in your intestines.

Now, what about wheat germ's potential to battle cholesterol? Frankly, the research has been limited, although promising. Most studies have been conducted in France, where volunteers with elevated cholesterol levels (between 236 and 372) added 20 grams (2.7 tablespoons) of raw or partially defatted wheat germ to their daily diet for a month; then they increased their intake of wheat germ even more, up to a total of 30 grams (4 tablespoons) a day. Their total cholesterol declined 8.7 percent on the raw wheat germ after the first month; but interestingly, after an additional 14 weeks consuming the higher amounts of wheat germ, the improvements in cholesterol slipped a little (to a 7.2 percent overall decrease), although they were still significant. The message here seems to be one that I've discussed earlier: While unsaturated fats are clearly better than saturated ones, too much of even the unsaturated variety can eventually begin to create problems.

By the way, when you're choosing foods to sprinkle your wheat germ on, look beyond just cold breakfast cereals or oatmeal. Give it a home atop salads, which will take on the nutty taste of the wheat germ without losing the flavors of the fruits and vegetables that are their basic ingredients. Or put a little in recipes for muffins and other baked foods. But, again, rather than tossing it on everything within sprinkling distance, don't forget that because wheat germ does contain fat (even though it's mostly unsaturated), you should eat it only in moderation.

■ ■ ■ ■

Chapter 7

Vitamins And Mineral To The Rescue: Antioxidants And Other Nutrients

For years, most mainstream nutritionists insisted that if you were eating a well-balanced diet, there was no need to "waste" money on vitamin and mineral tablets. While some diehards still cling to this party line, they're quickly becoming a minority, particularly when the goal is maximum well-being, including reducing the risk of heart disease and high cholesterol.

To begin with, many Americans simply don't consume a healthy diet. Most of us, in fact, lead fast-paced, eat-on-the-run lives, and unfortunately, our diets often suffer as a result. True, by adopting the 30-Day BLITZ, you'll probably be increasing the amounts of vitamin-rich fruits, vegetables and grains you eat, but still, I believe that supplements can help ensure that your diet is full of essential nutrients.

I also have some concerns with the Recommended Dietary Allowances (RDAs). Keep in mind that when these RDAs (or Daily Values, as they're now often called) were created, they were set at levels intended to prevent nutrient deficiencies -- not to achieve optimal health. Thus, when the RDA for vitamin C was established at 60 mg. for adults, that was the amount of C deemed necessary to prevent deficiency disorders like scurvy; but as subsequent research has demonstrated, much higher levels of vitamin C can play a role in preventing and/or reducing the severity of everything from the common cold to cardiovascular diseases to some types of cancer. In much the same way, there's a growing and persuasive body of evidence showing that, in many cases, your body can benefit from other vitamins and minerals at much higher doses than the RDAs.

So in creating the 30-Day Cholesterol BLITZ, I wasn't about to overlook the possible benefits of vitamin and mineral tablets. This vitamin and mineral terrain isn't fully charted yet, and ongoing research is going to give us much more information on which to base recommendations. However, considering what we already know, I urge you to press into service particularly those vitamins and minerals already shown to contribute to improved cholesterol and/or triglyceride levels and to better heart health. Your goal should be to take them in optimal -- but still safe -- doses.

In creating this component of the program, I felt that the best place to start was with a good multi-vitamin/mineral supplement which ensured that I was getting the RDAs -- and more -- of many essential vitamins and minerals. I've chosen a multi-vitamin without iron, because studies have shown that this mineral can possibly cause an increase in the oxidation of cholesterol -- which increases its likelihood of being laid down as plaques in your arteries. Of course, if you have anemia or excessive blood loss, you may have to take supplemental iron; ask your doctor. But if you don't *need* it, skip it. You may choose from the many good supplements on the retail shelves (see the contents of my multi-vitamin tablet below--it also contains a few minerals) or conveniently order them from our nutrition club (see page 119). I believe a tablet a day can provide you with a good foundation which you can build upon with other supplements.

CONTENTS OF MY MULTI-VITAMIN

Nutrient	Dose
Beta-Carotene (pro-vitamin A)	10,000 I.U.
Vitamin D	400 I.U.
Vitamin C	150 mg.
Natural Vitamin E (succinate)	100 I.U.
Vitamin B-1 (thiamine)	25 mg.
Vitamin B-2 (riboflavin)	25 mg.
Vitamin B-6 (pyridoxine)	25 mg.
Vitamin B-12 (cobalamin concentrate)	100 mcg.
Niacinamide	100 mg.
Pantothenic Acid	50 mg.
Biotin	300 mcg.
Folic Acid	400 mcg.
PABA	25 mg.
Choline Bitartrate	25 mg.
Inositol	25 mg.
Calcium (citrate & carbonate)	25 mg.
Magnesium (aspartate & oxide)	7.2 mg.
Potassium (aspartate & citrate)	5 mcg.
Zinc (from zinc picolinate)	15 mg.

Antioxidant vitamins

In the last few years, antioxidants -- namely, vitamins C, E and beta-carotene (a form of vitamin A) -- have received plenty of attention, and deservedly so. When it comes to heart disease, antioxidants have emerged as important allies, and in a number of ways. First of all, evidence shows that they wage a no-holds-barred battle against "free radicals", which are toxic chemical substances that can injure healthy cells. Unless these free radicals are inactivated, they can make you much more vulnerable to a number of chronic illnesses, including atherosclerosis and heart disease (as well as cataracts and colon, breast and lung cancer).

But there's even more to antioxidants, particularly when it comes to your cholesterol. Not only might they elevate your HDL (good) cholesterol level, but they interfere with the "oxidation" of LDL (bad) cholesterol, the process that leads to the depositing of LDL in the coronary arteries.

Just how persuasive is the evidence supporting the benefits of antioxidants? It's becoming strong enough to turn many people into "true believers" in the power of these nutrients (and I'm one of them). One study evaluated the benefits of various doses of vitamin C in nearly 700 men and women. Some of them took under 60 mg. of vitamin C per day; others consumed more than 180 mg. The individuals with the high vitamin C consumption had HDLs that were 11 percent greater than their low-C counterparts. Not only that, but their blood pressure readings were 8 percent lower.

In studies at the University of California, San Diego, researchers looked at the effects of vitamins C (2,000 mg. a day), E (1,600 mg.) and beta-carotene (60 mg.) -- taken alone and together -- upon cholesterol levels. More than any other antioxidant, vitamin E interfered with LDL's susceptibility to oxidation -- by 30 to 50 percent -- thus limiting the deposition of LDL in the artery walls. Without a doubt, however, much more research needs to be done, and not all recent studies have reached positive conclusions. Early in 1996, for example, two major studies concluded that supplements of beta-carotene do not prevent heart disease -- and might even increase the risk of lung cancer among smokers.

Still, taken as a whole, the evidence is still persuasive enough for me to incorporate the following antioxidant supplements into the BLITZ:

Vitamin E -- 400 IU a day

Beta Carotene -- equivalent to 25,000 IU of Vitamin A

Vitamin C -- 2,000 mg. a day (1,000 in the morning, 1,000 in the evening)

In creating these and other recommended dosages for cholesterol-fighting supplements, I've not only evaluated carefully what the studies have shown, but I've also scanned the vitamin shelves and looked at the doses that are commonly available. I've insisted upon keeping the recommended intakes at safe levels, too; some people, for instance, have found that starting immediately at a 2,000 mg. daily dose of vitamin C has caused transient side effects such as diarrhea, nausea, and headaches; for that reason, I advise building up your intake gradually, starting at 250 mg. a day, and then increasing that dosage by 250 mg. each month until you reach the 2,000 mg. level.

Calcium

Particularly if you're a woman with an eye out toward guarding against osteoporosis (a common bone-thinning disease), you might already be aware of the need for extra calcium in your diet. But calcium may do more than build strong bones; it might also provide protection against cardiovascular diseases, which is why I've incorporated calcium supplements into the BLITZ.

Some studies have even suggested that calcium has a direct cholesterol-lowering effect, but frankly, more research is needed before I can enthusiastically count on this mineral to join the front lines of the battle against cholesterol. Still, consider these findings: At the University of Minnesota, men and women with high cholesterol levels consumed 1,200 mg. of calcium a day, divided into 400 mg. doses taken with each meal. After six weeks, their LDL cholesterol dropped 4.4 percent -- a figure that may not sound especially impressive until you consider that it translates into a reduction in heart-attack risk of about 9 percent. At the same time, their HDL cholesterol levels increased 4.1 percent.

How does calcium work? One hypothesis is that it may cut cholesterol counts by interfering with the absorption of saturated fats, but this is just a theory at this point.

As research into the calcium-cholesterol link continues, the evidence is presently much more persuasive that calcium can decrease blood pressure readings. In many people, calcium appears to counteract the blood-pressure-raising effects of too much salt in the diet, probably by causing the excretion of greater amounts of sodium. A recent "meta-analysis" at Oregon Health Sciences University, evaluating all the available studies on this topic, concluded that calcium pills decreased blood pressure readings in 75 percent of patients; the declines averaged 5 to 7 points for systolic pressure (the upper figure), and 3 to 4 points for diastolic pressure (the lower one).

Those kinds of findings are good enough for me. My calcium-supplementation recommendation is to take a total of 700 mg. a day (500 mg. in the morning, and 200 mg. in the evening), although if you're also taking it to maintain bone strength, you may want to increase the dose; the National Osteoporosis Foundation advises a daily calcium intake of up to 1,500 mg. in postmenopausal women not taking estrogen replacement therapy.

Interestingly, the ordinary antacid, TUMS, is almost pure calcium. So whenever I have heartburn, I chew a few, knowing I'm helping build my calcium at the same time. Especially try the new "tropical fruit flavors"; they're delicious.

Are there any risks to taking too much calcium? Not really, unless you're prone to developing calcium-based kidney stones. Otherwise, except for occasional constipation, adverse effects of calcium supplementation are very rare. Nutrition experts advise if your daily dosage of calcium causes constipation, it can be prevented by increasing the daily dosage of magnesium. By the way, some studies have found that if you take your calcium pills with a meal, they can interfere with the absorption of iron in your diet; so I advise taking these supplements at other times of the day.

Niacin

Early in my own cholesterol-lowering campaign, my doctor put me on the prescription drug Mevacor. As important (and even life-saving) as this and similar drugs can be, once my cholesterol level had dropped, by the 30th day of my BLITZ, I stopped taking this medication with my doctor's blessing -- and instead introduced niacin to the program.

There's one very important point to keep in mind about niacin. If you are taking a cholesterol-lowering prescription drug (Mevacor, for example), talk with your doctor before self-prescribing niacin. Serious side effects may develop if you mix niacin with some of these drugs, so to be safe, consult with your physician first.

Niacin is part of the B-vitamin group (it's often called vitamin B3), and its cholesterol- lowering effects are undeniable -- not only in reducing LDL cholesterol but in increasing HDL levels as well. Even so, don't lose sight of this important caveat: Some studies showing the positive effects of niacin have used large doses (3,000 mg. a day), which are so high that your doctor should know you're taking it at those levels so he or she can monitor you for possible side effects.

I recommend a much lower dose -- 200 mg. of niacin supplements a day, divided into two 100 mg. doses. It still provides some cholesterol-cutting effect, while keeping the chances of side effects very low. However, to reduce the risks even further, I take non-flushing niacin, a special formulation that eliminates a common side effect of this vitamin -- namely, a flushing sensation and a reddening of the skin that is often called "hot flashes" by people who experience it; most often, it occurs an hour or two after taking the pill, and is caused by the release of histamine (a chemical that dilates the blood vessels) in the body.

By the way, there is also some evidence that if you take an aspirin tablet about 30 minutes before your niacin pill, you can avoid this flushing; but it's easier, I believe, just to look for the non-flushing version of niacin when you're shopping, and eliminate the risk of this unpleasant sensation altogether.

Chromium

In this era of nutritional hype, chromium hasn't received as much attention as calcium and other minerals. But if there's a shortage of chromium in your diet, don't be surprised if it's contributing to a spiraling cholesterol count.

Consider the findings of a study at Oklahoma State University, in which people took a 150-microgram capsule of chromium each day. After 12 weeks, the group as a whole did not show any changes in their cholesterol readings; but when researchers looked more closely at only the people whose cholesterol had been high (240 or above) when the study began, these individuals had experienced a decline in their total cholesterol of 12 percent, along with a 14 percent decrease in their LDL cholesterol.

Other research both in the U.S. and abroad has confirmed the positive role that chromium can play in improving your cholesterol. A study in Israel, for example, concluded that a 250-microgram daily supplement of chromium could elevate HDL cholesterol ("good" cholesterol) by 21 to 25 percent in men with heart disease, over a time period averaging 11 months. Studies have also shown that chromium deficiencies can help trigger adult-onset (type II) diabetes and an insensitivity to insulin.

Nevertheless, nothing comes easy. Chromium has been under fire recently, particularly in late 1995 when a study concluded that at very high doses (3,000 to 6,200 times the typical supplemental use!), chromium picolinate (a common form of chromium) injured chromosomes in human cells grown in the laboratory. Before you get too nervous, however, remember that this admittedly serious side effect occurred with extremely elevated doses. Chromium proponents, in fact, still insist that with normal intake levels, this mineral is important, health-promoting -- and safe.

Perhaps the best way to get your chromium is in food, such as broccoli, bran cereals, brewer's yeast and whole wheat. But even though there is no RDA for chromium, many researchers believe that the average American isn't getting enough of it in his or her diet. For its possible benefits in cholesterol control, I take chromium supplements -- 200 micrograms twice daily.

Folic acid

Sometimes called folate or folacin, folic acid is one of the B-complex vitamins, and for years doctors (particularly obstetricians) have been speaking up in support of its role in preventing neural tube defects (like spinal bifida) in developing fetuses. For that reason, folic acid is widely recommended for women of childbearing age. But more recently, researchers have also established a link between folic acid and heart disease prevention.

Here's the connection: A number of studies have found that folic acid, along with two other B vitamins (B6 and B12), can reduce blood levels of an amino acid called homocysteine. When your homocysteine levels are high, your coronary arteries are more prone to clogging, and for two reasons -- excessive homocysteine damages the interior lining of the artery walls; and it also increases the "stickiness" of blood components called platelets that assist in clotting. In a study of male health professionals at Harvard Medical School, those with the highest amounts of homocysteine were three times as likely to have a heart attack as those with the lowest levels.

Where can you find folic acid? Munch on foods like green leafy vegetables (such as spinach) and fruits (apples, oranges). In my own case, to ensure that I'm getting enough of this underappreciated vitamin, I also take it in supplement form. While the RDA for folic acid is 200 micrograms for men and 180 micrograms for women, many doctors believe it needs to be increased, and thus I take a higher dose -- 400 micrograms. Multivitamin tablets, by the way, usually contain folic acid, often at a 400-microgram level.

MY DAILY SUPPLEMENT GUIDE

After months of evaluating the research, Rainey and I have created the following supplement guide. I take supplements twice a day in the following doses (as you can see, in nearly all cases, I rely on amounts higher than the RDAs):

* First, I take one multi-vitamin/multi-mineral capsule per day: I use a variant without iron, since you don't need supplemental iron unless you're regularly losing blood. Iron has been incriminated in some research as an element which can increase the oxidation of cholesterol -- leading to more plaque clogging your arteries.) In addition to my multi-vitamin, I take the supplements shown in the following two charts:

MY MORNING SUPPLEMENT GUIDE

	My Dose	RDA
Morning		
Antioxidants:		
vitamin C	1,000 mg.	60 mg.
vitamin E	400 I.U.	12-15 I.U.
vitamin A	25,000 I.U.	5,000 I.U. (it's equivalent of beta-carotene)
potassium	92 mg.	--
zinc	50 mg.	12-15 mg.
folic acid	400 mg.	180-200 mg.
garlic capsule	400 mg.	--
Complex of two nutrients in one tablet:*		
calcium	500 mg.	800-1,200 mg.
magnesium	250 mg.	280-350 mg.

* *For some people it is better to take equal mg. amounts (i.e. 500 mg. calcium and 500 mg. magnesium) of calcium and magnesium because of the "tendency" of calcium to cause constipation. Adjust dosage to match your gut's type of motility.*

Complex of two nutrients in one tablet:		
chromium	200 mcg.	--
niacin (non-flush)	100 mg.	13-19 mg.
Fiber supplements:		
apple pectin capsules	1,000 mg.	--
guar gum powder (stirred in orange juice)	4 grams	--
oat bran capsules	1,500 mg.	--

MY EVENING SUPPLEMENT GUIDE

	My Dose	RDA
Evening		
vitamin C	1,000 mg.	60 mg.
calcium	200 mg.	800-1,200 mg.
magnesium	250 mg.	280-350 mg.
potassium	92 mg.	--
chromium	200 mcg.	--
niacin (non-flush)	100 mg.	13-19 mg.
Fiber supplements:		
apple pectin capsules	1,000 mg.	--
guar gum powder (stirred in orange juice)	4 grams	--
oat bran capsules	1,500 mg.	--

key: mg. = milligrams
mcg. = micrograms

Note: The above supplements were part of my original BLITZ and I've kept on with them daily. But in our continuing research, we've discovered some additional worthwhile additives. These include beta sitosterol complex (beta sitosterol-200 mg., campesterol-100 mg., and stigmasterol-80 mg.), gugulsterone-25 mg., and ginger root-550 mg. Our explorations are ongoing.

■ ■ ■ ■

Chapter 8

Excelling At Exercise: The Benefits Of Getting Physical

When I launched the Cholesterol BLITZ, I was, at best, a "light" exerciser. Although I wasn't the stereotypical "couch potato," I didn't lace up my walking shoes nearly as often as I should have -- perhaps once or twice in a good week. All that has changed, however. When the BLITZ began, I knew I wasn't training for the Boston Marathon, of course, but still I recognized that by becoming even modestly active, I could influence my cholesterol level and my risk of heart disease in a positive way. The evidence, in fact, is quite persuasive: If you want to shape up your cholesterol count, you need to get into shape yourself!

Here's what regular exercise can do. It can boost your HDL cholesterol, and protect your overall cardiovascular system by turning your heart into a more efficient pump. It can also activate an enzyme (called lipoprotein lipase) that transports triglycerides out of the bloodstream. Exercise helps control your blood pressure, keeps your bones strong, reduces your chances of developing adult-onset diabetes, and tones up your muscles. It can melt away stress, fight depression, and aid digestion. It can help you fight obesity, too, which is an important risk factor for heart disease and other serious illnesses. Not bad for a commitment of 30 minutes, a few times a week!

Now, just how strong is the evidence that exercise can raise HDL cholesterol? In a year-long study at Stanford University, researchers put a group of moderately overweight volunteers on a low-fat diet; half of them also began exercising (jogging, brisk walking) up to 45 minutes a day, three times a week. The HDL cholesterol levels of the men who exercised climbed an impressive 13 percent, compared to only a 2 percent increase among the sedentary group.

When I decided to incorporate exercise into the BLITZ, my goal was (and still is) to raise my heart rate for 30 minutes per session, without too much of the huffing and puffing that can make exercise seem more like work than fun. There are actually a lot of exercise options to choose from: brisk walking, swimming, dancing, bicycling, hiking, jogging, and racquetball, among many others. I chose walking, committing myself to four days a week, without fail.

But what if you're one of those people who is activity-phobic, and who just hates to exercise? All I can say is, "Give it a try." Like you, I used to cringe when I'd hear exercise fanatics proclaim, "Find an exercise you enjoy, and have fun doing it." Well, particularly in the beginning, I never felt in radiant bliss when I exercised, and I expect that you may not, either. I considered it a necessary "chore" that was important for my greater good. In fact, until I was deep into the BLITZ, I never liked exercise very much.

I suggest that you do whatever you have to do to make exercise a little more tolerable. I began by using a treadmill and watching TV as I walked; it really does make the time go faster. By the way, I also found that this treadmill walking is more efficient than my strolls through the neighborhood; the outdoor walks often turned into sauntering -- my speed would frequently slow to a snail's pace, sometimes coming to a complete halt if the scenery was distracting or I stopped to talk to someone.

Over the months, I have actually begun to look forward to working out (believe it or not!). My body feels so toned and fit that I really want to keep doing what makes that possible. Also, I've never really pushed myself to the point where exercise has seemed like punishment, leaving my body screaming for mercy. In fact, since I've built up gradually to my present level of activity, it has never really been a strain or struggle.

My first day on the treadmill, I walked at a speed of two miles per hour with the platform on a level plane; when I got tired, I slowed down to one mile per hour -- but even in those early days, I always made sure I stayed on the treadmill for 30 minutes. Today, I'm walking at a pace of four miles per hour at an incline of four percent grade. And I'm still going for 30 minutes a session.

Of course, there are days when I don't feel like working out. That happens to everyone from time to time. Right now, in fact, you probably have a dozen excuses on the tip of your tongue why you just can't exercise today. Maybe it's too cold (or too hot) outside … perhaps you woke up with the sniffles … or you might be expecting an important phone call at any minute. Do not let this kind of shortsighted self-talk stand in the way of your good health. I schedule my workout sessions just like I do any other appointment -- and I stick with it. If you feel your exercise sessions becoming monotonous, blot out the boredom by varying your physical activity -- perhaps using the treadmill two days a week, and dancing through an aerobics class on an additional two days. For some people, a little friendly persuasion works wonders: an exercise partner can keep you motivated to stick to your workout schedule.

By the way, I've also added upper-body weight training to my own exercise program, using an inexpensive set of free weights (I started with five-pound dumbbells, and have gradually increased the amount of weight). I've even begun using an all-purpose weight machine that Rainey and I discovered in the exercise room in our condominium building. This component of my workouts has given new definition to my arms and shoulders, and has enhanced my overall sense of well-being, even if it hasn't made a direct contribution to conquering my cholesterol problem. Picture the sedate scientist of my past life transforming to Rainey's new description of me: "A hunk, a hunk of burning love!"

The take-home message, then, is simply to get moving. You don't have to go to extremes to turn exercise into one of the most useful parts of this program.

One important caveat, however: Particularly if you've been inactive for many years, or have serious health problems (like heart disease, diabetes, arthritis, or obesity), see your doctor before you begin exercising. He or she may recommend that you undergo certain tests, and adopt a workout program designed specifically for your health situation.

One other warning: As you progressively get into shape, be prepared for compliments from family and acquaintances, along the lines of, "You look great. Are you exercising, or on a diet, or what?" And as you meet fellow exercisers, which inevitably happens, you'll get the warm, smug bonding that comes from knowing you're all "into" healthy activity -- instead of staying as the couch potato you were.

■ ■ ■ ■

Chapter 9

More Blast From The BLITZ: Taking Charge Of Your Cholesterol

Because the BLITZ is designed to attack high cholesterol from multiple directions at one time, I've added still other elements to the program. Each of them, in its own way, can help deflate your cholesterol count:

Relaxation techniques

Stress is an almost inevitable part of everyday life. Whether it's pressures at work, the frustrations of rush-hour traffic, or anxieties over balancing your checkbook, you may often feel as though you're in a state of high alert -- which can eventually take a toll on your cardiovascular health.

Here's what happens when stress becomes a constant companion. When you're feeling tense, your body releases stress hormones that surge through your bloodstream; like an overloaded fuse that eventually blows, the wear-and-tear of chronic stress can trigger sharp increases in blood pressure readings, as well as an accelerated buildup of plaque deposits in your coronary arteries. Some studies have shown that constant stress batters your cholesterol levels by increasing both total and LDL cholesterol counts.

Unfortunately, stressful life events may not be completely avoidable (how do you escape bumper-to-bumper traffic or anxiety over high crime rates?). Even so, you can learn to block stress' negative effects with relaxation techniques. One of the most effective was developed by Harvard professor Herbert Benson, M.D., and is described in detail in his book, The Relaxation Response. Or try this simple breathing exercise: Take a slow, deep breath, filling your abdomen and chest with air; then slowly exhale while visualizing yourself breathing out all your tension -- and repeat 6 to 10 times.

Massage

Another way to ease your stress is with massage therapy. The skin is the largest organ of our bodies and physical touch can be pleasant, nourishing -- and habit-forming. As part of the relaxation portion of the BLITZ, I've incorporated receipt of a weekly massage into the program, believing that, along with the deep-breathing techniques, it might be therapeutic for my entire cardiovascular well-being. A massage can dissolve away areas of bodily tension, and as it relieves feelings of stress, it might help bring high blood pressure (and perhaps even cholesterol readings) under control.

Weight reduction

Obesity is a risk factor for heart disease. If you're carrying too much weight, those excess pounds can boost your triglycerides and LDL cholesterol, and cut into your HDL cholesterol; by contrast, losing a few pounds can have the opposite effect.

For example, in a study at the University of Pennsylvania, men who lost an average of 24 pounds experienced five percent *increases* in their protective HDL readings. In another study, this one at the University of Maryland, researchers concluded that losing weight may be even more effective than aerobic exercise in helping overweight men reduce their chances of developing heart disease. A 10 percent weight loss, they discovered, was more likely than a three-times-a-week exercise plan to depress both cholesterol and blood pressure readings.

So what's the best way to shed your excess pounds? While fad diets seem to be a way of life for many Americans, you need to approach weight loss in a more sensible way to achieve permanent results. Try the BLITZ -- specifically, low-fat eating and regular exercise -- to make obesity a part of your past for good.

And you *will* lose weight on my BLITZ. I lost 14 pounds in the first 30 days -- decreasing from 195 to 181. And in the next 30 days, I lost 16 more. That's 30 pounds off in 60 days -- pretty good! Now I hover around 160 pounds. Most people notice the trimness. One drawback: I had to spend $100 having the waists of all my pants taken in -- but isn't that the kind of expense you love?

Stop smoking

There are a lot of reasons to crush out your cigarettes for keeps, and your heart health is one of them. The chemicals in cigarettes can injure the walls of your coronary arteries, making it easier for fatty deposits to accumulate there. Smoking can also undermine your HDL cholesterol count, reducing it by up to 15 percent, while elevating your LDL cholesterol and triglyceride levels.

Of course, anyone who has ever tried to stop smoking knows just how difficult it can be -- but millions of people have successfully quit (three million Americans a year, to be exact). If you need help, contact the local chapters of the American Cancer Society and the American Heart Association, which offer inexpensive smoking-cessation programs as well as pamphlets that can guide you toward giving up those cigarettes permanently.

Prescription Drugs

As I mentioned earlier, when my doctor saw just how high my cholesterol and triglyceride counts had become in my pre-BLITZ days, he prescribed a drug (Mevacor) to help defuse my elevated cholesterol quickly. No doubt about it, these cholesterol-busting and lipid-fighting drugs (and there are several of them) can be effective and even life-saving for many people.

If your own doctor prescribes medication, you need to keep taking it until he or she says it's OK to stop. But because my own blood tests demonstrated such significant improvements while on the BLITZ, I suggest that you have your cholesterol rechecked after 30 days on this program; if you experience the same kind of success I did, your physician may agree that you no longer need the help of drugs to keep your cholesterol in check. Or at the least, maybe your dosage can be reduced or you can be switched to a medication with less risk of side effects than the one you're currently taking.

Bear in mind, however, that because of genetics or other unknown factors, not everyone will respond to the BLITZ with such dramatic improvements as I did. I feel such individuals will be few in number, but for them anti-cholesterol drugs may continue to be appropriate -- and may prolong their lives. In a recent study published in the journal Lancet, Scandinavian researchers evaluated nearly 4,400 men and women (ages 35 to 70) who had angina (chest pains) or had suffered a heart attack, and who also had high cholesterol levels. Half of the individuals took cholesterol-reducing drugs; the others were given a placebo. After five years, the medication group not only experienced a plummeting of their LDL cholesterol (by an average of 35 percent), but they had a 42 percent lower coronary death rate compared to the patients on placebo.

Even if you're taking cholesterol-cutting medication, however, that doesn't give you carte blanche to live your life more recklessly than you might otherwise do. While on these drugs, you still need to eat right, stay physically active, and adopt the other recommendations in the BLITZ.

BETA-BLOCKERS: A TWO-EDGED SWORD

If you have hypertension, there are many effective prescription medications that can extinguish your high blood pressure readings. Some of the most popular drugs are in the beta-blocker family -- compounds such as atenolol (Tenormin), metoprolol (Lopressor), and propranolol (Inderal). These beta-blockers slow down the heart rate, which reduces the heart's blood output and in turn lowers blood pressure.

But although beta-blockers are marvels for controlling hypertension (as well as irregular heartbeats and angina pain), they may cause problems of their own. Beta-blockers can send triglycerides soaring in some patients, as well as suppress HDL levels, when taken in high doses. Many physicians are not aware of this. If you have elevated triglyceride levels and are on beta-blockers, talk to your doctor about whether it makes sense to switch to another class of antihypertensive drug for at least a while, to see if in the absence of the beta-blocker your triglyceride levels decline.

Get social support

In large part, I owe my own success on the 30-Day BLITZ to Rainey. Not only did she and I create this program as a team, but we have shopped, cooked, dined and exercised together.

I'm a strong believer that a support system of family and friends is invaluable when you need a shoulder to lean on, particularly when trying to make significant lifestyle changes like those in the BLITZ. As you begin this program, encourage your family to join you, eating in a more healthful way and agreeing to keep high-fat foods out of the refrigerator and cupboards where they can tempt you. Your family members can also share the cooking responsibilities when you try the recipes in this monograph, and they can become your exercise partners.

There's another support alternative available as well: Rainey and I have established a toll-free phone number that you can call 24 hours a day, where you'll hear motivational messages that also include the latest research findings on cholesterol-reduction. You are not alone, and when you feel as though you're "falling off the wagon" and need your spirits boosted, our 800 number (1-800-508-2582) is a place you can turn. This monograph is only the beginning of our communication with you. Whenever you need encouragement, reassurance or more information, please call. Or write us at: The BLITZ Dept., Advanced Health Institute, Suite 301-CB; 851 Fifth Avenue N., Naples, FL 33940.

■ ■ ■ ■

Chapter 10

Getting Started

It's finally time to take the first step in "leaving life as you once knew it." You now know what needs to be done; the challenge is to do it.

As you embark on this program, you might be feeling some anxiety. I know I did, fearing that I was about to be relegated to a culinary life devoid of variety and taste. But my transition went much more smoothly than I had imagined. I expected to feel deprived (of my favorite foods, for example), but I never did. I expected to feel hungry, but the bulkier high-fiber foods, fruits and vegetables kept me with a "full" feeling. There are many more healthful options than I first realized, and my taste buds have never felt mistreated.

To help keep you motivated, don't lose sight of just how rapid your improvements will probably be. In just 30 days, you can dramatically crush your high cholesterol count, and in the process, slash your risk of a heart attack and other serious health problems. My own cholesterol levels plummeted more than 200 points in just one month!

There's no need to become fanatical on the BLITZ (but it will probably help if you do!). If you have an occasional lapse -- perhaps overindulging in high-fat foods because you couldn't pull yourself away from the buffet table at a social event -- don't panic. Pick up the program where you left off and get back on track. If you're sticking with the plan most of the time, you'll keep making progress.

So be prepared for an exciting 30 days. If you're like Rainey and me, you will probably be so delighted with your improvements that you'll decide to make your new lifestyle a <u>permanent</u> part of your life. Frankly, that was not my original intent. My initial goal was to stick with it for 30 days, and then if my results were encouraging, I'd adopt a "modified" BLITZ -- letting some alcohol pour back into my life, enjoying a three-egg omelette now and then, and feasting on an occasional juicy hamburger.

But my progress after those first 30 days was so astounding that I bought into the BLITZ for good -- and with continuing positive results. After my second month on the program, my triglycerides had fallen an additional 16.5 percent (down to 207), my cholesterol level was still a safe 182, and my weight had plummeted to 165 pounds (an overall decline of 30 pounds). My blood pressure remained low, too, at 124/82 -- and that's without any medication. No wonder my attitude has remained so upbeat!

I believe this program can lead you down the path to a lifelong commitment to healthier living. Sure, Rainey and I splurge a little bit on birthdays and other special occasions. But we feel so healthy and energetic that there's no turning back for us. The BLITZ may lower your cholesterol for good, too, and tilt the odds in your favor for a longer, more productive life.

Please start right away. You're on the brink of a transformation toward wellness and greater happiness. I wish you good health -- not only for the next 30 days of the Cholesterol BLITZ, but for many years to come.

YOUR BLITZ ACTION PLAN

Easy steps to start you on your way to a healthy heart:

1. Clean out your cupboards at home and rid yourself of the temptation to eat foods high in saturated fat.
2. Flip to the back of this report and order your Blitz vitamin starter kit.
3. Set an appointment with your physician to discuss this program and your current health condition. Get your cholesterol measured.
4. Tear out the shopping list in the Appendix and purchase a supply of the suggested healthy foods.
5. Eat low-fat healthy foods one-day-at-a-time starting today. Now you know what and why you should eat certain foods.
6. Commit yourself to perform some form of exercise today, even if it is only a 10 minute walk. (Increase it by 5 minutes every day and you will be walking for 45 minutes by the end of your first week.)
7. Tell someone about your plan and seek his/her support to help you stay on it.
8. When eating out, choose the restaurant carefully. Make it easy on yourself and choose places where you can get foods that taste good but which are healthy choices.
9. Adjust to a new mind set for social occasions. As a priority, enjoy the social side of the event first, and the food served at the event second.
10. For extra motivation, record your dramatic, effortless weight loss week by week.

REMEMBER THE "NUGGETS" OF THE BLITZ

1. Switch to low-fat and non-fat eating whenever possible.
2. Eat a largely vegetarian diet.
3. Fill up with fiber.
4. Replace fat and cholesterol-packed foods with Near 'Nuff substitutes.
5. Choose breakfast foods designed for cholesterol reduction.
6. Eat healthy snacks, such as fresh fruit and raw vegetables.
7. Select healthier desserts.
8. Choose your condiments carefully.
9. Avoid alcohol.
10. Take advantage of other cardiovascular protectors.
11. Read food labels.
12. Add supplements, vitamins and minerals to your daily diet.
13. Exercise at least three times every week.
14. Learn to relax.
15. Lose excess weight. (This comes effortlessly when you incorporate several of the BLITZ tactics all at the same time.)
16. Stop smoking.
17. Get social support and encouragement.

KEEP UP WITH THE LATEST PROGRESS

Our intensive research is continuing, and we're constantly finding new items worthy of trial or permanent inclusion. We'll keep you abreast of this progress through our newsletter. For a free, no-obligation copy, send a request to the address below.

The BLITZ Dept.
Advanced Health Institute
851 Fifth Avenue North, Suite 301-CB
Naples, FL 33940

In recent months, Rainey and I have created many recipes for the 30-Day BLITZ. We have discovered that we don't have to sacrifice most of the familiar tastes that make eating so pleasurable, nor swear off many of our favorite dishes (although we have adapted some of them in low-fat ways).

The box below will provide you with general food preparation tips to help keep your cooking low in fat. If you're not using the right techniques in the kitchen, you can turn a healthy food into a dish that sends your cholesterol soaring. We've used these same guidelines in preparing the recipes that follow.

COOKING THE LOW-FAT WAY

- Use methods that allow the fat to drip off during cooking. These include broiling, baking, braising and roasting. Stay away from frying, which actually increases the amount of fat you'll consume, in part by absorbing some of the oil used to coat the cooking pan.
- The longer you cook meat, the greater the amount of fat that will drip off and be lost. Thus, well-done meat is less fatty than rare meat.
- Cook vegetables by steaming or stir-frying. If you boil your vegetables instead, they'll lose many of their vitamins and minerals. Stir-frying veggies, however, will retain more of their nutrients and texture. Microwaving is another way of keeping the loss of nutrients to a minimum.
- When cooking with oil, use it very sparingly, and choose oils high in monounsaturated or polyunsaturated fats (such as canola, olive, safflower, sunflower, corn and soybean oils). Avoid oils rich in saturated fats (like coconut and palm oils). No-stick cookware and non-stick cooking spray are great alternatives to bathing your food in oil or butter.
- Use skim milk (evaporated skim milk tastes rich and creamy) instead of whole milk in cooking, such as when preparing baked goods.
- If you eat beef, make the portion size less than the size of a deck of cards, choose the leanest cuts and trim away the visible fat before cooking.
- Remove the skin from chicken before cooking.
- Steam or poach fish to help keep it as low in fat as possible.

Recipes

Page Guide

■ BREAKFAST IDEAS

Oat Bran Muffins

2 Cups Arrowhead Mills Oat Bran Pancake Mix
½ tsp. NoSalt
4 tbsp. sugar
Optional: ¼ cup chopped pecans or walnuts (Remember, these contain lots of fat.)
Optional: ¼ cup raisins
1 cup evaporated non-fat skim milk
3 tbsp. canola oil
3 egg whites "beaten stiff"

Preheat oven to 425 degrees and line a 12 space muffin tin with paper cupcake liners. (This makes cleanup a breeze.) Mix together the dry ingredients. Then, combine with milk, canola oil and the three egg whites. Mix all ingredients quickly with a few strokes of a mixing spoon and then pour mixture into the paper cups, filling them almost to the top. Bake 15 to 20 minutes or until toothpick inserted in muffin will pull out clean.

Servings: Makes 12 to 18 muffins.

Oat Bran Pancakes

1 cup Arrowhead Mills Oat Bran Pancake Mix
1 cup evaporated non-fat skim milk
2 - 3 egg whites
1 tbsp. canola oil
½ tsp. NoSalt

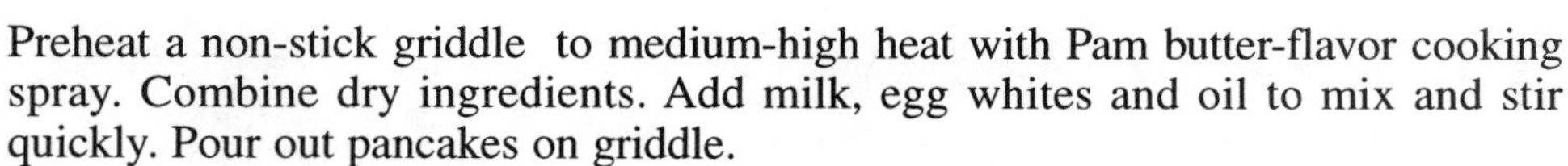

Preheat a non-stick griddle to medium-high heat with Pam butter-flavor cooking spray. Combine dry ingredients. Add milk, egg whites and oil to mix and stir quickly. Pour out pancakes on griddle.

If you wish to have fruit in your pancakes, sprinkle blueberries or thin slices of bananas into the mix as it cooks on its first side, then flip the pancake to complete cooking process. (This way you can prepare your pancakes "to order" according to each family member's preferences.)

Serve with no fat butter spray and a sugar free, fat free syrup. This also tastes great with a side of Morning Star fake "bacon" strips.

Servings: Makes 8 to 10 Pancakes.

APPETIZERS

"How to Catch a Man" Mushrooms

1 tbsp. olive oil
2 tsp. chopped garlic
juice of 1 lemon
¼ cup low sodium soy sauce
1 tsp. of Tabasco or Cajun Sunshine
1 package button mushrooms cleaned
½ tsp. coarse ground black pepper

In non-stick pan, saute garlic until it's brown. Then, add lemon juice, soy sauce, Tabasco, button mushrooms and ground black pepper. Combine all ingredients and saute mushrooms in sauce on medium high, while continually coating them with the sauce until tender. Serve piping hot.

Servings: 4 - 6

Humus

1 can garbanzo beans
1 tsp. olive oil
1 tsp. tahini
2 tbsp. sesame seeds
1 tsp. balsamic vinegar

Blend all ingredients in food processor until smooth and well blended. Serve with pita bread or fat free crackers.

Servings: 4 - 6

Tabouli (Bulgar wheat)

1 package of Tabouli
2 cups cherry tomatoes
1 cucumber
½ small onion
3 tbsp. fresh parsley
juice of ½ lemon
1 tbsp. olive oil

Prepare Tabouli according to package directions. Chop all of the tomatoes, cucumber, onion and parsley. Mix all ingredients and chill for 30 minutes. Serve with pita bread.

Servings: 4 - 6

Perfect Party Dip

1 16 oz carton of fat free sour cream
1 cup chopped fresh cauliflower
1 tbsp. horseradish
3 tbsp. fresh dill
½ cup no-fat or low-fat Swiss cheese, cut in tiny pieces
1 tsp. of Tabasco
1 tsp. of Worcestershire

Mix all ingredients together and chill for 30 minutes. Serve as a party dip with low-fat crackers or place on bed of salad greens with sliced tomatoes and cucumbers for a super summer lunch.

Servings: 4 - 6

■ SOUPS

Healthy Gazpacho (Spanish-style Chilled Vegetable Soup)

3 medium ripe tomatoes
1 large European cucumber
½ each red, yellow, green and seeded stemmed bell pepper
2 ribs celery
2 small cloves of garlic
1 32 oz. can low-sodium V-8 juice
optional: 1 small fresh jalapeno pepper

Chop tomatoes, cucumber, bell peppers and celery and place in bowl. Mince finely and add garlic cloves and jalapeno. Stir in low-sodium V-8 juice with chopped veggies -- start with 16 oz. Adjust to personal preference of thinness.
Add Tabasco sauce to your taste. Chill in refrigerator for 30 minutes to an hour before serving. (This tastes great the next day, too.) Garnish with fresh chopped parsley and a dollop of no-fat sour cream.

Servings: 4 - 6

Cold Fruit Soup

1 cup fresh Strawberries
3 fresh peeled seeded plums
1 peach
1 tsp. lemon juice
3 tsp. artificial sweetener

In a medium saucepan, cook all ingredients over low heat for 15 minutes. Then put them in food processor for quick turn to chop and mix. Pour in bowl, add artificial sweetener and chill for 1 hour before serving.

This soup can be made at any time of year using a combination of 3 "in season" fruits.

Servings: 4 - 6

Omma's Crock Pot Cabbage Soup

Head of one cabbage, cut in small pieces
2 medium onions chopped
3 potatoes chopped
3 carrots chopped

2 cans Healthy Choice Low-Fat Chicken Broth
NoSalt and Season Pepper
Bay leaf
Dash of garlic salt
Dash of Maggi

Place all ingredients in crock pot. Cook on high for 1 hour, then turn to low setting for 6 to 8 hours longer.

Servings: 4 - 6

Corn Chowder

3 medium potatoes, diced
1 can no-fat chicken broth
½ cup water
6 ears fresh sweet corn,
 cut off the cob
1 medium onion, diced
½ cup skim milk
Two 8 oz. cans non-fat evap. skim milk
¼ of red bell pepper - seeded,
 stemmed and diced
3 ribs of celery, diced

In a medium saucepan, boil potatoes in broth and water for 20 minutes. Remove cooked potatoes from liquid and set aside. To same pot, add remaining vegetables, NoSalt and pepper -- cook until done, about 20 minutes. Using food processor, puree potatoes. Combine milks with vegetables in the saucepan. Stir in pureed potatoes and simmer 10 minutes longer while flavors blend. Garnish with chopped parsley.

Servings: 4 - 6

Cucumber Dill Soup

2 large European Cucumbers - washed, but unpeeled
12 oz non-fat sour cream
⅓ cup fresh dill
½ cup skim milk
Juice from ½ lemon
2 garlic cloves
NoSalt, fresh ground pepper to taste

Place cucumbers, dill, lemon, garlic, and seasonings in blender or food processor and blend well until smooth. Pour mixture into medium bowl and gently stir in non-fat sour cream and skim milk. Refrigerate for 30 minutes. Serve with tiny sprig of dill and non-fat garlic croutons.

Servings: 4 - 6

■ SALADS

South of the Border Black Bean Salad

1 large can black beans, drained and rinsed
2 ribs celery, diced
1 medium tomato, diced
½ scallion, diced
1 can chopped green chilies
1 can mexicorn, drained
2 tbsp. cilantro
juice of 2 limes
1 tsp. garlic, minced
NoSalt and season pepper to taste
optional: 2 tbsp. prepared salsa

Combine all ingredients in large bowl and chill for 1 hour. This salad improves with time, so it's even better the next day.

Cucumber Salad

6 sliced fresh gherkins -- peeled and sliced (small cucumbers sold in produce section)
3 small sweet onions peeled and sliced- separate rings
1 cup white vinegar
1 cup water
NoSalt and Pepper to taste and artificial sweetener

Combine and chill.

Servings: 4 - 6

No-Fat Cold Slaw

1 head of cabbage shredded
1 medium onion chopped
3 ribs of celery chopped
½ each red and green bell pepper seeded and stemmed -- chopped
1 tsp. celery seed
2 tsp. artificial sweetener
⅔ cup water
1 cup vinegar
Nosalt and pepper to taste

Mix all dry chopped ingredients, including celery salt, and set aside. Combine water, vinegar, salt and pepper and artificial sweetener and bring to a boil. Immediately pour this over all mixed ingredients. Chill.

Servings: 4 - 6

Kashi

1 package of Kashi (a grain)
one cucumber diced
one small sweet onion diced
two ribs of celery diced
juice of 2 lemons and one lime
½ cup chopped fresh peppermint
NoSalt and pepper to taste

Prepare Kashi according to directions; let cool. Add remaining ingredients and refrigerate for 30 minutes before serving.

Servings: 4

Green Beans & New Potato Salad

Dressing:
¼ cup olive oil
½ cup balsamic vinegar
¼ cup water
¼ tsp. paprika
1 clove garlic, crushed
¼ crushed basil
NoSalt or Mrs. Dash

Salad ingredients:
2 medium tomatoes, chopped
1 medium European cucumber sliced
6 radishes thinly sliced
shredded salad greens
10 very small new red potatoes
¾ lb fresh green beans

Mix ingredients for dressing, shake and set in refrigerator. Trim stems on washed green beans and "snap" into 1-inch pieces. Steam until tender (about 20 minutes). Remove from heat and allow to cool. Simultaneously wash and boil potatoes until tender (about 20 minutes). In a large bowl add all salad ingredients and refrigerate for 1 hour. Toss dressing gently on ingredients and serve.

Servings: 4 - 6

Fabulous Fiber Salad

Salad:
½ of 8 oz can garbanzo beans, drained
1 8-oz can red kidney beans, drained
favorite salad greens, in bite size pieces
1 medium tomato, chopped
½ each green and yellow bell pepper (stemmed and seeded), chopped
1 medium European cucumber chopped
2 Tbsp. fresh onion (chopped very fine)

Combine all ingredients and refrigerate for 1 hour.

Dressing:

½ cup non-fat plain yogurt
2 Tbsp. chopped fresh dill
Juice of ½ lemon
Nosalt "Season salts and pepper" to taste
1 tsp. of olive oil (optional)

Combine all ingredients for dressing and toss with salad ingredients to coat all veggies.

Servings: 4

Absolutely No-Fat -- but "Tastes Great" Salad

Cut into irregular bite-size chunks:
4 farm fresh tomatoes (important to get the best you can buy)
1 green bell pepper (seeded and stemmed) cut into ¼ inch thin strips - 1 inch in length
½ of sweet onion sliced paper thin (making half rings)
NoSalt and pepper seasons to taste.

Prepare veggies and place in bowl to marinate for 1 hour.

The combination of flavors with the natural juice or "dressing" of the tomatoes makes this simple combination a real winner for the no-fat taste buds.

Serves: 4

■ SANDWICHES

Mediterranean Mouthfuls

¼ inch round slices of fresh eggplant (use ½eggplant with skin on)
1 sweet red pepper, stemmed and seeded and cut in 1 inch wide lengths
2 Portabella mushrooms Sliced in ½ inch widths
Liquid Smoke bottled seasoning
8 slices (½ inch thick; 2 per sandwich) of fat-free whole grain loaf
(8 inch whole round loaf)
romaine lettuce leaves washed
fresh tomatoes sliced
favorite mustard
I Can't Believe It's Not Butter! spray

Use outside grill or hot griddle coated with olive oil-flavored Pam and brown both sides of vegetables about 5 minutes each. Baste with Liquid Smoke generously on each side of veggie. Spray both sides of bread with no-fat butter spray, place on grill or griddle and toast until golden brown. Pile up the smoky flavor veggies on toasted bread, spread with mustard, add romaine lettuce leaf and tomato slices. Cut in half for serving -- it is more than a mouthful! Serve with lentil soup.

Servings: 4 sandwiches - 1 per person

Guilt-Free Lox and Bagels

4 plain or onion bagels
1 package Philadelphia fat-free cream cheese
1 tbsp. finely chopped onion
½ tsp. chopped fresh dill
1 tsp. capers
6 oz left-over poached salmon

Mix cream cheese with onion, dill, and capers. Spread on toasted bagel and top with salmon.

Servings: 4

"Unbelievable BLTE" Low-fat version of America's favorite

2 sliced tomatoes and shredded lettuce
2 slices of whole grain low-fat bread
Morning Star Breakfast Strips (veggie bacon)-2 slices per sandwich
2 egg whites per sandwich
I Can't Believe It's Not Butter! spray
Nayonaise

"Fry" veggie bacon and egg whites (2 @ a time per sandwich) on non-stick griddle coated with Pam Butter Spray. Spray no-fat butter spray on 2 slices of bread, then toast on griddle until golden brown. Build BLTE: layer cooked egg, then 2 slices of cooked bacon on one slice of bread. (You could also add a slice of Kraft non-fat sharp cheddar cheese to melt on egg -- very yummy.) Then pile on lettuce and tomato and toasted second slice to top of your sandwich.

Serve with Simply Potatoes hash browns (Cook with non-stick butter spray -- no need to add oil.) Cook on side of your griddle while preparing the BLTE's.

Servings: 1 sandwich per person

"Say Cheese" Veggie Burger

Answer your cravings to a burger with the great variety of frozen veggie burgers. See Near 'Nuff list on page 25 for selection to look for at your grocery store.

Pan fry veggie burgers in non-stick pan coated with Pam spray or grill on the barbecue, about 6 minutes per side of burger. Melt a slice of non-fat extra sharp cheddar cheese over it. Load your burger with onion, tomato, lettuce and mustard onto a sesame bun.

Serve with B&M 99% Fat Free Brick Oven Baked Beans.

Servings: 1 sandwich per person

The Blue Plate Veggie Sandwich

1 sliced cucumber
1 sliced tomato ½ avocado (Easy does it -- only 2 slices, as it's very high in fat.)
lots of sprouts
no-fat grated cheese
no-fat whole grain bread
mustard

Layer ingredients on toasted bread and spread on a tsp. of dijon mustard.

Servings: 2 sandwiches (serves one person)

■ MAIN COURSE

Vegetarian Barley Chili

1 cup chopped onion
1 medium size eggplant,
  chopped into ½ inch cubes (skin- on)
1 cup chopped celery
1 cup chopped bell pepper
2 toes of garlic, minced finely

Spray non-stick Dutch oven sauce pan with olive oil-flavored Pam cooking spray and, on medium heat, saute veggies down to wilted stage. Eggplant should take on a brownish look with a consistency similar to ground beef.

1 large can whole tomatoes cut into fourths
2 small cans low sodium tomato sauce
1 small can low sodium tomato paste -- using the same can add 2 cans of water
2 bay leaves
Add ¼ tsp. of Vegit, Spike and Mrs. Dash seasonings
Add ½ tsp. of chili powder
1 can red kidney beans
1 cup cooked barley (use package directions)
Add the tomatoes, tomato paste sauce and seasonings to the sauted veggies in the Dutch oven and cook on low simmer for 30 to 45 minutes. Add kidney beans, and the precooked barley to the chili mixture and heat through for another 5 minutes.

Garnish with no-fat formagg® grated cheddar cheese or fresh chopped onion, or both.

Serve with a mixed green salad, as this is truly a one-dish meal.

Servings: 4 - 6

Vegetable Oriental Stir Fry

1 onion chopped
1 zucchini chopped
1 red and green bell pepper (stemmed and seeded) chopped
2 cups Chinese Cabbage shredded
1 yellow squash chopped
1 minced garlic clove
1 cup low sodium soy sauce
2 tsp. olive oil
2 tbsp. cornstarch
1 cup cooked brown rice

Spray wok or non-stick fry pan with Pam olive oil spray. Cook all vegetables quickly on medium-high heat for about 5 minutes. Add 1 cup of low sodium soy sauce, cornstarch for thickening. Toss gently until vegetables coated and sauce is smooth and free of lumps.

Serve immediately over precooked rice. Serve a fresh fruit salad on the side of fresh cut cantaloupe pieces and blueberries.

Servings: 4

Aunt Bessie's Louisiana Red Beans and Rice

(Crock Pot version)

1 package dried red beans
1 onion chopped
2 ribs of celery chopped
2 toes of garlic minced fine
2 bay leaves
NoSalt, pepper seasonings
2 packages Goya ham flavored seasonings
1 tbsp. olive oil
1 cup cooked brown rice

Wash and soak 1 pound package of dried red beans overnight. Early the next day, place chopped onion, celery, and garlic in non stick pan sprayed with Pam olive oil flavored spray and saute for 10 minutes until wilted. In crock pot, add beans (rinse and throw away soaking water), wilted veggies, 3 cups of water, bay leaves, ham flavored packages, olive oil and various no salt seasonings to Crock Pot. Cook on high for 1 hour and then turn to low setting to cook 6 to 8 hours longer.

Serve red beans piping hot over freshly cooked rice with no-fat cole slaw.

Servings: 4 - 6

Ratatouille Cous Cous Casserole

1 large eggplant, chopped
1 medium green bell pepper, chopped
2 medium zucchini, chopped
1 large onion, chopped
1 ½ tsp. finely chopped fresh garlic
¼ cup white wine
3 medium ripe tomatoes or 1 can 16 oz drained stewed tomatoes, chopped
½ tsp. each dried basil and thyme
1 bay leaf
NoSalt, pepper seasonings and garlic powder to taste
4 tbsp. chopped fresh parsley
⅓ cup freshly grated Parmesan cheese
1 package pre-cooked Cous Cous

Pour ¼ cup white wine in skillet, saute onions, bell pepper and garlic until onion is transparent. Add eggplant and zucchini, saute until tender -- about 6 minutes. Add the tomatoes and seasonings. Add all remaining ingredients except Parmesan cheese. Cover and continue cooking about 10 minutes.

Place cooked Cous Cous in deep baking dish, cover with ratatouille (cooked vegetables) add cheese, and bake for 10 minutes.

Serve hard rolls and salad for a complete meal.

Servings: 4 - 6

Garlic Chicken

4 chicken breasts -- deboned and skinned -- cut in half
2 cups white wine
30 unpeeled cloves of garlic
NoSalt and pepper seasonings

Place chicken in ungreased baking pan and cover with seasonings. Pour white wine over chicken and place garlic cloves in and around chicken pieces. Cover tightly with aluminum foil. Bake at 325 degrees in oven for 1 hour and 15 minutes.

Garnish serving plates with cooked garlic -- great to spread on bread or eat right out of the skin.

Servings: 4 - 6

Poached Salmon

1 salmon fillet (about 1 to 1½ pounds)
1 whole lemon sliced
1 cup champagne
1 carrot, sliced
3 sprigs fresh dill
1 onion quartered
NoSalt, white pepper

Place salmon in dish with tight cover or use heavy foil large enough to seal tightly for steaming. Place lemon and dill decoratively on top of salmon, place carrot and onion around sides. Pour champagne over salmon and seal tightly so that no steam can escape. Bake at 375 degrees for 20 minutes. Note: This time varies depending on the thickness of the salmon. It is very important not to overcook the salmon and to remove it as soon as it turns opaque.

Chill left-overs; they make a great lunch when served with no fat sour cream.

Servings: 4 - 6

Poached Grouper

4 Grouper fillets
½ cup white wine
juice from ½ lemon
diced onion
1 cup prepared salsa
¼ cup capers
Optional: 1 tbsp. chopped black olives

Saute onion in white wine and lemon juice until onions are transparent. Add grouper and salsa; cover tightly. Cook over medium heat about 10 minutes or until done. Add capers and chopped black olives.

Serve with cooked Brown Basmati Rice and Black Beans.

Servings: 4

Oven "Tastes Fried" Chicken

1 cup no-fat yogurt
1 tbsp. Worcestershire sauce
NoSalt, Mrs. Dash,
pepper, paprika to taste
1 tsp. onion powder
1 cup crushed corn flakes
⅓ cup whole wheat flour
½ tsp. sage and ½ tsp. paprika
6 boneless skinless chicken thighs

Remove skin and any fat from chicken pieces, rinse, and dry. Roll in yogurt, then cornflake mixture and place on non-stick baking pan. Bake at 400 degrees for 30 to 40 minutes.

This recipe can also be used to cook small fish fillets, simply substitute sage for lemon pepper.

Servings: 6

Stuffed Peppers

2 Green, 2 Red, and 2 Yellow med. bell peppers (washed, seeded, & hollowed out)
1 can black beans, drained
1 medium onion diced
1 cup cooked Kashi (a grain)
¼ cup fresh mushrooms
½ tsp. minced garlic
NoSalt and season pepper to taste

Except the hollowed out peppers, mix together ingredients above. Spoon mix into the peppers. Place stuffed peppers in pyrex baking dish. Then, prepare sauce with the following ingredients:

Sauce:
1 can Campbell's Healthy Request Cream of tomato soup
1 cup of Pritikin Fat-Free Marinara Sauce
2 Tbsp. Ketchup
½ cup water
Mix liquid ingredients, pour over peppers, seal dish tightly with foil or lid. Bake at 375 degrees for 30 minutes.

Servings: 4 - 6

"Any Time" Healthy Breakfast Quiche

1 package Simply Potatoes shredded hashbrowns
1 medium onion, diced
2 tbsp. olive oil
NoSalt and pepper to taste
½ cup chopped mushrooms
2 tbsp. chopped bell pepper
1 package frozen chopped spinach, thawed
9 egg whites, beaten stiff
1 tsp. Worcestershire sauce
dash of Tabasco sauce
1 tsp. whole grain mustard
grated no-fat cheese

Brown potatoes and onions in oil and place in 9½ x 11 baking dish. Set aside. Saute mushrooms and bell pepper for 5 minutes or until tender and stir in spinach until heated through. Add veggie mixture to beaten eggs in mixing bowl. Mix until well blended. Then, add water, Worcestershire, Tabasco, mustard and seasonings to mixture and pour over potatoes. Bake at 375 degrees in oven for 30 minutes or until done. Add no-fat cheddar cheese and place under broiler two to three minutes longer just prior to serving.

Serve with salsa or baked tomatoes.
Servings: 4 - 6

Pasta Fagoli

5 garlic cloves, chopped
1½ tbsp. olive oil
6 medium peeled, diced tomatoes
2 tsp. fresh, chopped rosemary
4 cups no-fat vegetable bouillon (Swanson's)
4 oz. tomato sauce
red pepper flakes to taste
NoSalt and pepper
1 can red kidney beans, drained
1 can garbanzo beans, drained
1 can white beans, drained
1 package Farfalle (or you may use bow tie rotelle or elbow pasta instead)

Lightly saute garlic in deep fry pan with olive oil until light brown. Add tomatoes and rosemary and cook over medium heat for another 6 minutes. Add all remaining ingredients and cook until pasta is done about 10 minutes more.
Servings: 4 - 6

Enchilada Chicken Casserole

1 cup plain non-fat yogurt
2 tbsp. chopped green chilies
¾ tsp. chili powder
½ tsp. ground coriander
1 tsp. minced garlic
3 cups cooked chicken or turkey
12 low fat pre-packaged tortillas (Bake for 5 minutes prior to assembly)

Mix together all ingredients except tortillas. Arrange pre-baked tortillas in a shallow baking dish, alternate chicken mixture and more tortillas ending with a layer of tortillas. Bake for 20 minutes in 350 degree oven, then top with non-fat grated cheddar cheese and bake 10 minutes longer. Serve with salsa.

Grandma's Low-fat Chicken Pot Pie

6 small, boneless, skinless chicken breasts cut into small pieces
2 ribs of celery, chopped
1 medium onion, chopped
1 can no-fat chicken broth
1 can of water
1 carrot, peeled and sliced thinly
2 potatoes, peeled and chopped small
1 can Campbell's Healthy Choice 98% fat-free Cream of Celery soup
1 cup low-fat evaporated skim milk
1 cup low-fat Bisquick mix

Place chicken, celery, onion, broth, water, carrot and potatoes in deep sauce pan and bring to low simmer for 20 minutes. Stir in soup, then pour mixture into 9 x 9 Pyrex baking dish. In separate bowl, mix together milk and Bisquick until well blended, then pour on top of chicken mixture. Bake at 350 degrees in preheated oven for 30 minutes, until golden brown.

Servings: 4 - 6

Advanced Health Institute
851 Fifth Avenue North, Suite 301
Naples, FL 33940
800-508-2582 • Fax 941-261-6713

Thank you for your recent order. May we present Dr. X's Breakthrough:

The Doctor's 30-Day Cholesterol BLITZ

The BLITZ approach to cutting cholesterol offers new hope for you and your loved ones. The author is Dr. Leslie Norins, a noted physician, researcher and publisher in the health care industry. He successfully battled his high cholesterol and high triglyceride problem and slashed his blood lipid levels over 50% in just 30 days.

You're invited to read and try his unique multi-pronged approach to controlling cholesterol. And you're guaranteed to get the results you long for--or get a prompt refund. You now have hope from Dr. Norins' inspiring story and his practical answers for getting a healthy heart. We care about you and wish you the best.

Sincerely,

Rainey Norins

Rainey Norins
President

P.S. Can we send a gift copy to a friend or loved one? Share this healthy information with those in your life who need it most. Call 1-800-508-2582 and our friendly service representatives will be happy to help you.

Turkey Burgers

1 lb ground turkey
1 egg white
finely chopped onion
garlic
Catalina Fat-Free French Dressing
½ cup bread crumbs
Liquid Smoke

Mix together all ingredients and shape into 6 patties. Grill over hot coals or broil in oven and baste both sides with Liquid Smoke. Serve with "Catch a Man" mushrooms smothered on the top.

Servings: 4 - 6

Esther's Vegetable Pasta

1 lb. broccoli rabe cut in small pieces
1 lb. fresh asparagus cut in 1 inch pieces
2 toes of minced garlic
3 tsp. of olive oil
1 can stewed tomatoes
NoSalt and peppers
1 package angel hair pasta
fresh Parmesan cheese

Steam: broccoli rabe and fresh asparagus. Set aside. Saute garlic in olive oil about 5 minutes, until golden brown. Add 1 large can stewed tomatoes (whole or chopped). Add seasonings and previously cooked veggies. Toss with freshly cooked angel hair pasta. Serve immediately with a light shaving of Parmesan.

Servings: 4 - 6

Blitz Seal of Approval "Fettucini Alfredo"

2 cans no-fat evaporated skim milk
2 egg whites, beaten until frothy
3 tsp. olive oil
3 tsp. flour
1 package Farmer's low-fat cheese
1 package fettucini noodles, cooked
3 tbsp. fresh chopped parsley
fresh grated hard Parmesan cheese

Scald the skimmed evaporated milk. Heat olive oil and flour in separate pan and add scalded milk slowly. Alternate beaten egg whites and milk, while stirring constantly and briskly. Stir over low heat until smoothly blended. Stir in 1 small package of low-fat cheese until blended and just heated through. Toss immediately with fresh cooked fettucini (according to package directions) and sprinkle a small amount of fresh chopped parsley on top as garnish.

Servings: 4 - 6

Macaroni and Cheese Bake

1 package elbow macaroni
1 package of Farmer's low-fat grated cheese (or Breakstone's made with skim milk)
1 package fat-free cheddar cheese
6 egg whites, beaten
NoSalt, season peppers
Pam cooking spray
approximately 2 cups skim milk

Cook package of elbow macaroni per package directions. Drain pasta, place in large mixing bowl. Add package of Farmer's grated cheese. Add beaten egg whites, fat-free cheddar cheese and seasonings. Mix well. Spray casserole dish with Pam spray. Pour all ingredients into casserole dish and fill with skim milk within ½ inch of top edge of baking dish. Bake 1 hour in a 350 degree pre-heated oven.

Servings: 4 - 6

Louisiana Seafood Gumbo

(Rainey is from "Cajun country" and for the BLITZ she modified her family's recipe.)

2 tbsp. olive oil
2 tbsp. flour
2 onions, chopped
1 cup green onions, chopped
2 toes garlic, chopped
1 green pepper, chopped
1 cup of celery, chopped
2 cups okra
1 28 oz. can of whole tomatoes
1 tbsp. Oyster Sauce
NoSalt
½tsp. Red Pepper
4 bay leaves
1 lb. of peeled, deveined fresh raw shrimp
1 lb. can of fresh crabmeat
Optional: 1 dozen small, fresh oysters

Make a roux with oil and flour by mixing them over medium heat in a deep saucepan, stirring constantly until a rich brown color. Set aside. Coat non-stick pan with Pam olive oil-flavored spray and cook onions, garlic, green pepper, celery, until wilted -- about 10 minutes -- then add frozen okra and cook until it starts to "goop up". Add tomatoes and other seasonings with about a quart of water (The mixture should look more like soup than stew at this point.) Bring mixture to a medium boil for twenty-five minutes. Add shrimp and cook for 5 minutes longer. Add crabmeat and oysters. Crabmeat is already cooked and oysters will be "curled" or cooked through by the time it gets to the table.

Serve over cooked brown Basmati rice in bowls (pasta bowls make a good substitute for gumbo bowls) Toast crusty french bread that can be "sopped" or dipped into the juice of the gumbo instead of using any butter or margarine. Add a fruit salad for a complete meal suitable for the most discerning guests. (They won't guess it is low-fat, so don't tell!) File' (pronounced "fee lay") powder is used to add a little thickening to the finished gumbo, but if it is unavailable it will not be missed in your gumbo concoction.
Servings: 6 - 8

Bouillabaisse "Spicy Gulf Stew"

¼ cup olive oil
½ cup carrot, diced
½ cup celery, diced
1 tbsp. Garlic minced
3 medium tomatoes, peeled and chopped
1 cup dry white wine
1 cup clam juice
2 cups Swanson no-fat chicken broth
½ cup fresh lemon juice
2 bay leaves
2 tsp. parsley, chopped
1 to 2 tsp. Tabasco
NoSalt
fresh cracked black pepper
1 pound shrimp, in shells
1 pound scallops
1 pound firm fish grouper or scrod
1 pound mussels, in shells (in season!)

Add olive oil to sauce pan and heat. Add chopped vegetables, except tomatoes, to olive oil and saute until wilted, about 10 minutes. Add liquid ingredients, and spices, except Tabasco. Reduce heat and cook on low about 20 minutes more. Add Tabasco and tomatoes. Cook 5 minutes on low. Bring to a simmer. Add mussels and fish. Cook approximately 10 minutes on medium heat. Add shrimp and scallops. Cook 5 more minutes or until done, but do not overcook.

Any seafood in season may be substituted for those listed.

Servings: 4 - 6

Crock Pot Black Bean Soup with Rice

1 pound dry black beans, sorted and rinsed
1 can 14½ oz. chopped tomatoes
1 onion chopped fine
1 cup chopped celery
1 cup chopped green pepper
2 gloves garlic, minced
2 packages Goya Ham Seasoning
1 bay leaf
2 cans Swanson's no fat Chicken Broth
2 cups water
1 tbsp. olive oil
NoSalt and season pepper
1 cup of Basmati brown rice, cooked according to package directions
Non-fat grated cheddar cheese
Non-fat sour cream
Picante Sauce

Cover beans with water in a bowl and soak overnight. Rinse beans and place in crock pot with onions, tomatoes, green pepper, seasonings broth, olive oil and water. Cook on high for 1 hour, reduce heat to low and cook 6 to 8 hours longer. Remove and discard bay leaf. Take 2 to 3 cups of cooked beans and puree in blender or food processor and return soup mixture to pot. Add seasoning NoSalt and pepper.

Serve soup over cooked Basmati rice in bowl then garnish it with a dollop each of non-fat cheese, sour cream and Picante sauce.

Servings: 4 - 6

Vegetarian Lasagna

1 8 oz box lasagna noodles
2 tbsp. olive oil
1 large onion, chopped
3 garlic cloves minced fine
dash red pepper flakes
2 tsp. oregano
3 stalks celery, chopped
1 large carrot, chopped
1 bell pepper, chopped
1 half eggplant, chopped
1 29 oz. can crushed tomatoes
1 8 oz. Can tomato sauce
½ lb. mushrooms sliced
hard Parmesan cheese

Chop onion and saute in olive oil for 3 minutes. Add garlic and seasonings. Stir 1 to 2 minutes over medium heat. Add celery, carrot, bell pepper and eggplant and cook 8 minutes longer. Add tomatoes and sauce and mushrooms and heat through. Cook noodles al dente and then drain. Spray Pam on 9 x 12 oven casserole dish and layer noodles, sauce, and a small shaving of hard Parmesan cheese for three layers. Bake at 350 degrees for 1 hour.

Servings: 4 - 6

■ SIDE DISHES

Nature's Buttered Scalloped Potatoes

6 Yukon Gold medium potatoes, peeled, washed and sliced in ⅛ inch crosswise slices
1 - 2 cans fat-free evaporated skim milk
3 tbsp. chopped onion
3 tbsp. of flour
NoSalt and pepper seasonings

Preheat oven to 325 degrees. Spray baking dish with butter-flavored Pam. Layer dish with slices of potatoes, seasonings, chopped onion and 1 tbsp. of flour on each layer. Repeat for 3 layers. Pour skim evaporated milk over potatoes until milk shows through potatoes on top layer. Cover dish and bake at 325 degrees for ½ hour, then uncover and bake for 1 additional hour.
(Ukon Gold Potatoes have a natural butter flavor; you will not miss butter omitted -- trust me.)

Servings: 4 - 6

Garlic and Celery Mashed Potatoes

6 Yukon Gold Potatoes, peeled and cubed
1 cup water
3 cloves of garlic peeled and chopped

3 ribs of celery cleaned and cut into 1 inch lengths
2 Cans no-fat chicken stock
½ cup Pet evaporated fat-free skim milk
NoSalt and pepper to taste

Combine and bring all the ingredients (except milk) to a medium boil and cook until tender, about 30 minutes. Remove from heat and add ½ cup of evaporated skim milk. Whip potato mixture with electric mixer until smooth and creamy. When serving, heap mounds of potatoes on individual plates, and spray top of mound with no-fat I Can't Believe It's Not Butter! spray for a "no calorie" extra burst of butter flavor.
Servings: 4 - 6

Garlic Stuffed Onion

4 large Spanish onions
4 slices Morning Star Breakfast Strips imitation bacon strips, browned and crumbled (or bacon bits)
1 whole head garlic, peeled and chopped
1 can water chestnuts, drained and chopped
¼ cup bread crumbs
½ tsp. Nosalt
¼ tsp. black pepper
1 cup apple cider

Put unskinned onions in boiling water for 8 minutes, remove and cool, peel and scoop out insides, leaving 3 layers of onion to form shell. Set layers aside. Chop the cooked onion. Mix together bacon, chopped onion, garlic, water chestnuts, bread crumbs and seasonings. Fill each onion with the mixture, place in casserole. Pour apple cider over tops of filled onions. Bake at 375 degrees for 45 to 50 minutes.

Servings: 4

Baked Onion

4 large sweet onions
(Similar to Maui or Vadalia)
4 beef bouillon cubes
season pepper
about 2 cups of red wine

Peel each onion and cut top and bottom off so that one surface is flat. Cut a V or funnel hole in top of onion, being careful not to cut through the bottom. Discard the cut-out piece. Place onion on double layer aluminum foil large enough to completely wrap up like a package. In the hole you cut place one beef bouillon cube, sprinkle onion with seasoned pepper, and fill hole with red wine. Repeat for each onion. Seal onion with aluminum foil so that no steam escapes. Place in baking dish and bake at 350 degrees for 1 hour.

Servings: 4

Crunchy Potatoes

1 lb potatoes
Pam cooking spray
medium onion, finely chopped
1 tbsp. olive oil

Clean and cube 1 lb of potatoes. Boil until tender. Spray non-stick skillet with Pam spray. Finely chop a medium onion and saute onions until clear. Add cooked potatoes and cook on low heat with 1 tbsp. olive oil for 30 minutes, until potatoes are brown and crispy.

Servings: 4 - 6

Red Bliss Potato Salad

1 pound small red bliss potatoes, washed
3 ribs of celery, finely chopped
4 hard boiled egg whites, finely chopped
2 tbsp. of fresh dill
⅔ cup of Nayonaise
NoSalt and pepper seasonings

Boil potatoes for 20 to 30 minutes. Remove from water, drain, place in mixing bowl and cut into bite size chunks. Add celery, hard boiled egg whites, fresh dill, and seasonings. Mix all ingredients and chill.

Servings: 4 - 6

"Barely Believable" Fat-Free Butter Nut Squash

3 butter nut squash, washed and cut in half
1 bottle Catalina Fat-Free French Dressing
1 can whole cranberries
1 package Lipton Onion Soup Mix

Preheat oven to 375 degrees. Mix dressing, cranberries, and soup mix until blended and fill halves of squash with mixture. Bake for 45 minutes or until tender. Sauce mixture is great to use on baked boneless skinless chicken too.

Servings: 4 - 6

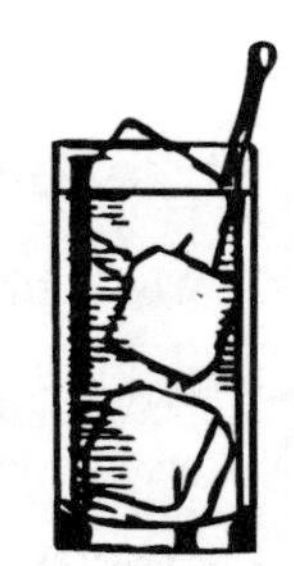

■ BEVERAGES

"Blow Your Head Off" Cocktail

(Leslie's Special Non-alcoholic Concoction)

½ bottle low sodium V-8 juice
2 ribs of celery
½ tsp. of horseradish
1 tbsp. Worcestershire sauce
1 tbsp. lemon juice
5 to 10 drops of Tabasco sauce (If you have just experienced a "bad day at black rock", go for the extra powerful kick of hot spices to calm your nerves.)
Several ice cubes

Place all ingredients in blender and mix until smooth. Serve.

Leslie's After Dinner Chocolate Soda

Version 1:
Pour a small amount of cold skim milk in the bottom of a tall glass. Then pour, from about 10 inches above the glass, one chilled can of Canfield's Chocolate Soda. Pour slowly to let the carbonation bubble up and mix the soda well with the milk.

Version 2:
Place in blender the following ingredients:
1 can of Canfield's Chocolate Soda (icy cold from refrigerator)
½ cup skim milk
2 tsp. Hershey's Cocoa (unsweetened)
2 to 3 package artificial sweetener
several ice cubes
2 tsp. Guar Gum powder

Blend all ingredients in blender except Guar Gum until creamy smooth. Add Guar Gum and blend for another 15 seconds at high speed. Pour in tall old fashioned malt or soda glass.

And voila, you have Near 'Nuff Chocolate "ice cream" Soda.

If you want even more "richness," before blending, add 1 small scoop frozen no-fat vanilla yogurt or frozen no-fat "ice cream substitute." If desired, top with squirt of no-fat "whipped cream" substitute from aerosol can.

Enjoy!

Appendix

With a lot of research and careful label-reading, Rainey and I have created a shopping list of brands and products that are compatible with The Doctor's 30-Day Cholesterol BLITZ. Look over this list before your next visit to the supermarket, or make a copy of it that you can take to the store with you. By filling up your shopping cart with these items, you'll give yourself a faster start toward reducing your cholesterol count.

SHOPPING LIST

The first step to lowering your cholesterol is to go to the grocery store, health food store, and a farmer's market to stock up on the items you will need to have at home, in your kitchen, and in your refrigerator to prepare healthy low-fat meals and snacks. Use this list as a guide and inspiration to gather the weapons needed to put your own BLITZ into action.

Fresh Produce

Vegetables and fruits are nature's fat free wonders when they are fully vine ripened and *fresh* (with the exception of avocados). Do yourself a favor and find the very best produce your area provides. When you replace fat in your diet with great flavor, it can be such a treat for your taste buds that it replaces the craving for the slippery fat taste of your former eating plan. Check out a "Farmer's" Market, "Street" Produce Market vendor, or a superb grocery store where you can select great tasting vegetables and fruits. If you hesitate paying $3 for some really great tomatoes, just think back to your past when you stopped in at Ben and Jerry's and paid $3 for a double scoop, which clogged your arteries.

What to buy:

Tomatoes -- the finest, freshest you can buy
Sweet Onions and Garlic
Green Onions
Celery (2 to 3 stalks of the smaller hearts of celery -- some for cooking, and plenty for eating raw every single day)
Carrots (The little packages of cleaned, ready to eat baby carrots taste great and are easy to munch!)
Red, Green and Yellow Bell Peppers
Eggplants
Zucchini
Yellow squash
Cucumbers
Portabella mushrooms
Shitake mushrooms
Button mushrooms
Fresh Herbs: cilantro, dill, basil, chives, rosemary, arugula, and parsley
Potatoes: Red bliss, Yukon gold, sweet and baking potatoes
Salad Greens -- Endive, Romaine, Spinach, Boston Bibb, Radicchio, and Ice Berg)
Cabbage

Broccoli and Broccoli Rabe
Sprouts
Asparagus
Corn on the cob
Green Beans
(Another rule of thumb is: If its green, buy it.)
Fresh Fruits: Bananas, Blueberries, Oranges, Apples, Cantaloupe, and Strawberries.

Your Local Health Food Store

What to buy:
Nayonaise (mayonaise substitute)
Veggie Hot dogs (in the frozen food section)
Arrowhead Mills Oat Bran Pancake Mix
Gimme Lean Meatless Ground Beef

Supermarket

What to buy:
Healthy low-fat or fat-free suggestions to try:
Dairy Section

Butter Substitutes:
Nabisco's Fleischmann's Fat Free Low Calorie Spread (5calories)
I Can't Believe It's Not Butter! spray

Cream or Milk Substitutes:
Skim Milk

No Fat Sour Cream Substitutes:
Naturally Yours Fat Free Sour Cream
Land of Lakes Fat Free Sour Cream
Breakstone Fat Free Sour Cream

No Fat Yogurt:
Dannon No Fat Plain Yogurt
Dannon Light (Fat Free flavored yogurt)

Cottage Cheese:
Light 'N Lively Fat Free Cottage Cheese
Breakstone Fat Free Cottage Cheese

Other Cheese Substitutes:
Philadelphia Fat Free Cream Cheese
Polly-O Mozzarella
Polly-O Ricotta

Cheese Slices:
Kraft Fat Free American, Swiss, and Sharp Cheddar

formagg®
Alpine Lace
Healthy Choice American
Borden Swiss, Sharp Cheddar
(Note: Many of the fat-free and low-fat cheeses are available in grated from also.)

Egg Substitutes:

Egg whites from the whole egg (discard the yellow)
Second Nature Eggs
Better n Eggs
Egg Beaters

Refrigerated Foods:

Simply Potatoes
Oscar Mayer Fat Free Bologna, Ham, Hot Dogs, Beef Franks, and Turkey Franks
Horseradish
Mission fat Free Tortillas

Salad Dressings:

Kraft Fat Free Salad Dressings: Ranch, Blue Cheese, and Honey Dijon
Wishbone Fat Free Italian
Alessi Balsamic Vinegar (aged)
Alessi White Balsamic Vinegar
Progresso Red wine vinegar
First Cold Pressed Virgin Olive Oil

Cereals:

Cold Cereals:

Health Valley: Real Oat Bran, 100% Natural Bran Cereal, and Healthy Fiber Flakes, Ralston Multi-Bran Chex, Nabisco Shredded Wheat 'Bran, Kellogg's All-Bran Extra Fiber, Crispix & Common Sense Oat Bran, Post Grape Nuts

Hot Cereals:

Quaker Oat Bran
Quaker Oatmeal

Beverages:

Crystal Light Mixes: Lemonade, Iced Tea, and Fruit Punch
Canfield Chocolate Sodas
Low-Sodium V-8 Juice
Alcohol Free Beer: O'Doul's
Alcohol Free Wine: Ariel or Sutter Home Fre'
Herbal Teas
Coffee

Soups, Staples and Canned Goods:

Pam Spray flavors: Olive Oil, Butter, Olive Oil & Garlic flavors
Carnation Light Evaporated Milk
Pet Evaporated Skimmed Milk
Tuna in spring water
B&M 99% Fat Free Brick Oven Baked Beans (vegetarian)
Campbell's Healthy Request Cream of Chicken, Tomato, and Cream of Mushroom
Campbell's Reduced fat Cream of Celery
Swanson's fat free chicken broth
Swanson's fat free vegetable broth
Baxter's Fat free onion soup
Dried Beans: Red beans, Black beans, and Lentils
Cache River Brown Basmati Rice
Carey's No-Fat Sugar Free Syrup (for pancakes)

Frozen Foods:

Morning Star Breakfast Strips (fake bacon)
Veggie Burger choices:
Morning Star Farms Meatless Garden Vege Patties:
Grillers, Veggie Burgers (plain and black bean), Better 'n Burgers
The Better Burger (veggie burger)
Garden Mexi Vegetarian Burgers
The Original Garden Burger
Oat Bran Bagels
Frozen Vegetables: Black Eyed Peas, Tiny Green Peas, Lima Beans and Extra Sweet Corn on the cob (when you can't get it fresh)
Haagen Dazs Chocolate Sorbet Bars

Condiments & Seasonings:

Morton Salt Substitute
NoSalt
Vegit All Purpose Seasoning
Mrs. Dash (salt free) Fine Ground Herbs & Spices
Mrs. Dash Extra Spicy (salt free)
Spike (salt free seasoning)
McCormick All Purpose Table Shake
McCormick All Purpose Seasoning
McCormick Lemon Pepper
McCormick Garlic and Herb
lemon herb seasoning
Creole seasoning
chicken and beef bouillon cubes or powder
Goya Ham Seasoning (powder)
Maggi seasoning
Worcestershire Sauce
Tabasco
Tiger Sauce

Pickapeppa Sauce
French's Mustard and variety of Dijon or hot Mustards

Desserts and Snacks
Jell-O Fat Free Sugar Free Pudding: chocolate and vanilla
fat free pretzels
popcorn (air pop it)
Guiltless Gourmet fat free tortilla chips
Guiltless Gourmet fat free dips and salsa
Louise's Fat Free Potato Chips
Crackers:
Health Valley Fat Free Crackers (Whole Wheat, Pizza, and Fire Crackers)
Nabisco Harvest Crisps: 5-Grain, Garden Vegetable, and Italian Herb.

The Blitz Nutrition Club

Order a 30-Day starter kit by mail from our "Nutrition Club" for fast, easy home delivery of all the hard-to-find vitamins and supplements in this program. Call Advanced Health Institute at 1-800-508-2582 to receive your free catalog.

What to buy:
An all purpose Multi-Vitamin
Antioxidants:
vitamin C ....1,000 mg.
vitamin E..400 I.U.
Vitamin A..25,000 I.U.
potassium ....92 mg.
zinc..50 mg.
folic acid ....400 mg.
garlic capsule ....400 mg.
Tablet-complex of two:
calcium 500 mg./magnesium............250 mg.
Tablet-complex of two:
chromium...200 mcg./niacin(non-flush) 100 mg.
Fiber supplements:
apple pectin capsules ....1000 mg.
guar gum powder
oat bran capsules................................1,500 mg.

Optional new discoveries to try are: beta sitosterol complex, gugulsterone, and ginger root.
[Our explorations are ongoing. Keep up to date with our monthly newsletter ***"Nature's Cures"*** *for the latest information to stay healthy and fit.* ***(800-508-2582)****]*

Index